Foundation
transition

for **AQA, Edexcel** and **OCR two-tier GCSE mathematics**

The School Mathematics Project

Writing and editing for this edition   John Ling, Paul Scruton, Susan Shilton, Heather West
SMP design and administration   Melanie Bull, Pam Keetch, Nicky Lake, Cathy Syred, Ann White

The following people contributed to the original edition of SMP Interact for GCSE.

Benjamin Alldred	David Cassell	Spencer Instone	Susan Shilton
Juliette Baldwin	Ian Edney	Pamela Leon	Caroline Starkey
Simon Baxter	Stephen Feller	John Ling	Liz Stewart
Gill Beeney	Rosemary Flower	Carole Martin	Biff Vernon
Roger Beeney	John Gardiner	Lorna Mulhern	Jo Waddingham
Roger Bentote	Colin Goldsmith	Mary Pardoe	Nigel Webb
Sue Briggs	Bob Hartman	Paul Scruton	Heather West

CAMBRIDGE UNIVERSITY PRESS
Cambridge, New York, Melbourne, Madrid, Cape Town, Singapore, São Paulo

Cambridge University Press
The Edinburgh Building, Cambridge CB2 2RU, UK

www.cambridge.org
Information on this title: www.cambridge.org/9780521689960

© The School Mathematics Project 2006

First published 2006

Printed in the United Kingdom at the University Press, Cambridge

A catalogue record for this publication is available from the British Library

ISBN-13   978-0-521-68996-0 paperback
ISBN-10   0-521-68996-1 paperback

Typesetting and technical illustrations by The School Mathematics Project
Other illustrations by Robert Calow, Steve Lach and Sam Thompson at Eikon Illustration
Cover design by Angela Ashton
Cover image by Jim Wehtje/Photodisc Green/Getty Images

The authors and publisher thank the following for supplying photographs: page 13 © ImageState/Alamy; page 15 © Guinness World Records 2002; page 23 © Ben Wood/CORBIS; page 48 © f1 online/Alamy; pages 53, 54, 89 (girl), 91, 96, 147 Paul Scruton; page 89 © Topham Picturepoint (small man, tall man in eastern clothing and tall woman), © Bettmann/CORBIS (tall man in suit); page 108 © Edd Westmacott/Alamy (skating), © Jeff Gynane/Alamy (London Eye); page 122 © Jeremy Hardie/zeta/Corbis; page 153 © CORBIS

The authors and publisher are grateful to the following examination boards for permission to reproduce questions from past examination papers, identified in the text as follows.
AQA   Assessment and Qualifications Alliance
Edexcel   Edexcel Limited
OCR   Oxford, Cambridge and RSA Examinations
WJEC   Welsh Joint Education Committee
The authors, and not the examination boards, are responsible for the method and accuracy of the answers to examination questions given; these may not necessarily constitute the only possible solutions.

The map on page 130 is based on mapping supplied by The Automobile Association © The Automobile Association Developments Limited 2006 LIC017/06 A03154 © Crown copyright. All rights reserved. Licence number 399221

NOTICE TO TEACHERS
It is illegal to reproduce any part of this work in material form (including photocopying and electronic storage) except under the following circumstances:
(i)   where you are abiding by a licence granted to your school or institution by the Copyright Licensing Agency;
(ii) where no such licence exists, or where you wish to exceed the terms of a licence, and you have gained the written permission of Cambridge University Press;
(iii) where you are allowed to reproduce without permission under the provisions of Chapter 3 of the Copyright, Designs and Patents Act 1988, which covers, for example, the reproduction of short passages within certain types of educational anthology and reproduction for the purposes of setting examination questions.

Using this book

This book, *Foundation transition*, is for students who have followed any 'support' course in key stage 3. It prepares up to GCSE grade F and can be used to cover or revise basic topics before students start *Foundation 1*, the first of the two main books for the Foundation tier.

To help users identify material that can be omitted by some students – or just dipped into for revision or to check competence – chapter sections estimated to be at national curriculum level 3 or 4 are marked as such. These levels are also given in the detailed contents list on the next few pages.

At the end of the contents list is a precedence diagram to help those who want to use chapters selectively or in a different order from that of the book.

Each chapter begins with a summary of what it covers and ends with a self-assessment section ('Test yourself').

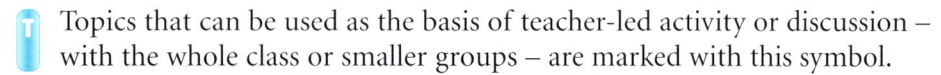

Topics that can be used as the basis of teacher-led activity or discussion – with the whole class or smaller groups – are marked with this symbol.

There are clear worked examples – and past exam questions, labelled by board, to give the student an idea of the style and standard that may be expected, and to build confidence.

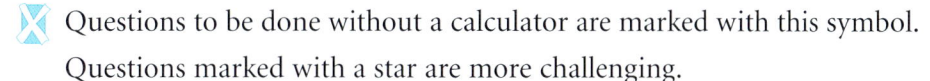

Questions to be done without a calculator are marked with this symbol.

Questions marked with a star are more challenging.

After every few chapters there is a review section containing a mixture of questions on previous work.

The resource sheets linked to this book can be downloaded in PDF format from www.smpmaths.org.uk and may be printed out for use within the institution purchasing this book.

Practice booklets

There is a practice booklet for each students' book. The practice booklet follows the structure of the students' book, making it easy to organise extra practice, homework and revision. The practice booklets do not contain answers; these can be downloaded in PDF format from www.smpmaths.org.uk

Contents

1 **Odds, evens, multiples, factors** 8
  A Odd and even numbers level 3 8
  B Divisibility level 3 9
  C Multiples level 4 10
  D Factors level 4 11

2 **Mental methods 1** 12
  A Place value level 4 12
  B Rounding level 4 13

3 **Shapes** 15
  A Circles level 3 15
  B Triangles level 4 16
  C Quadrilaterals level 4 18
  D More than four edges level 4 20
  E Shading squares to give reflection symmetry 22

4 **Adding and subtracting whole numbers** 24
  A Using written and mental methods level 3 24

5 **Listing** 26
  A Arrangements level 4 26
  B Combined choices level 4 27

6 **Multiplying and dividing whole numbers** 29
  A Multiplying whole numbers level 4 29
  B Dividing whole numbers level 4 30

Review 1 31

7 **Representing data** 32
  A Frequency charts and mode (for types of things) level 4 32
  B Dual bar charts 34
  C Pictograms level 4 35
  D Frequency charts and mode (for quantities) level 4 36

8 **Fractions** 39
  A Recognising fractions level 4 39
  B Finding a fraction of a number 40

9 **Decimal places** 42
  A One decimal place level 4 42
  B Two decimal places level 4 44
  C Decimal lengths level 4 46

10 **Median and range** 48
  A Median (for an odd number of data items) 48
  B Median (for any number of data items) 49
  C Range 50
  D Comparing two sets of data 51

11 **Mental methods 2** 53
  A Multiplying by 10, 100, 1000 53
  B Dividing by 10, 100, 1000 54
  C Multiplying by numbers ending in zeros 54

12 **Solids, nets and views** 56
  A Solids and nets level 4 56
  B Views 58

13 **Weighing** 60
  A Grams and kilograms 60
  B Using decimals 61

Review 2 62

14 **Time and travel** 63
  A Understanding 12-hour and 24-hour clock time level 3 63
  B Time intervals level 3 64
  C Working out starting times level 3 65
  D Timetables level 4 66

15 Angle 68
- A Drawing, measuring and sorting angles level 4 68
- B Turning level 4 71
- C Angles in shapes 72
- D Estimating angles 73

16 Length 75
- A Centimetres and millimetres 75
- B Metres and centimetres 76
- C Kilometres 77

17 Squares and square roots 78
- A Square numbers and square roots level 4 78
- B Using shorthand 79
- C Using a calculator 80

18 Adding and subtracting decimals 81
- A Adding decimals level 4 81
- B Subtracting decimals level 4 82
- C Mixed questions level 4 83

19 Mental methods 3 84
- A Multiplying and dividing by 4 and 5 level 4 84

Review 3 85

20 Number patterns 86
- A Simple patterns level 4 86
- B Further patterns 87

21 Estimating and scales 89
- A Estimating lengths 89
- B Reading and estimating from scales 92
- C Decimal scales 93

22 Multiplying and dividing decimals 96
- A Multiplying a decimal by a whole number 96
- B Dividing a decimal by a whole number 97

23 Area and perimeter 98
- A Shapes on a grid of centimetre squares level 4 98
- B Area of rectangle and right-angled triangle 100
- C Area of a shape made from simpler shapes 102
- D Using decimals 104

24 Probability 108
- A The probability scale 108
- B Equally likely outcomes 110

Review 4 113

25 Enlargement 114
- A Enlargement on squared paper 114
- B Scale factor 115

26 Negative numbers 117
- A Putting temperatures in order 117
- B Temperature changes 119
- C Negative coordinates 120

27 Mean 122
- A Finding the mean of a data set 122
- B Comparing two sets of data 124

28 Starting equations 126
- A Arrow diagrams level 3 126
- B Think of a number level 3 127
- C Number puzzles level 4 128
- D Using letters 128

29 Finding your way 130
- A Using a town plan level 4 130

Review 5 133

30 Volume 134
- A Counting cubes 134

continues >

31 Evaluating expressions 136
- A Simple substitution 136
- B Rules for calculation 137
- C Substituting into linear expressions 138
- D Formulas in words 140
- E Formulas without words 142

32 Estimating and calculating with money 144
- A Solving problems without a calculator level 4 144
- B Estimating answers level 4 145
- C Using a calculator level 4 146

33 Capacity 148
- A Litres, millilitres and other metric units 148

34 Drawing and using graphs 150
- A Tables and graphs 150
  (linear relationships)
- B Graphs and rules 154

Review 6 158

35 Fractions, decimals and percentages 159
- A Fraction and percentage equivalents level 4 159
- B Fraction and decimal equivalents 161
- C Fraction, decimal and percentage equivalents 162
- D Percentages of a quantity, mentally 163

36 Two-way tables 165
- A Reading tables level 3 165
- B Distance tables level 4 166

37 Scale drawings 168
- A Simple scales 168
  (1 cm represents n units)
- B Harder scales 171
  (2 cm and 5 cm represents 1 metre)

38 Using percentages 173
- A Percentage bars 173
- B Interpreting pie charts 174
- C Drawing pie charts 176
- D Comparing 177

39 Conversion graphs 179
- A Using a conversion graph 179
- B Drawing a conversion graph 181

Review 7 183

Answers 184

Index 221

The precedence diagram opposite, showing all the chapters, is designed to help with planning, especially where the teacher wishes to select from the material to meet the needs of particular students or to use chapters in a different order from that of the book. A blue line connecting two chapters indicates that, to a significant extent, working on the later chapter requires competence with topics dealt with in the earlier one.

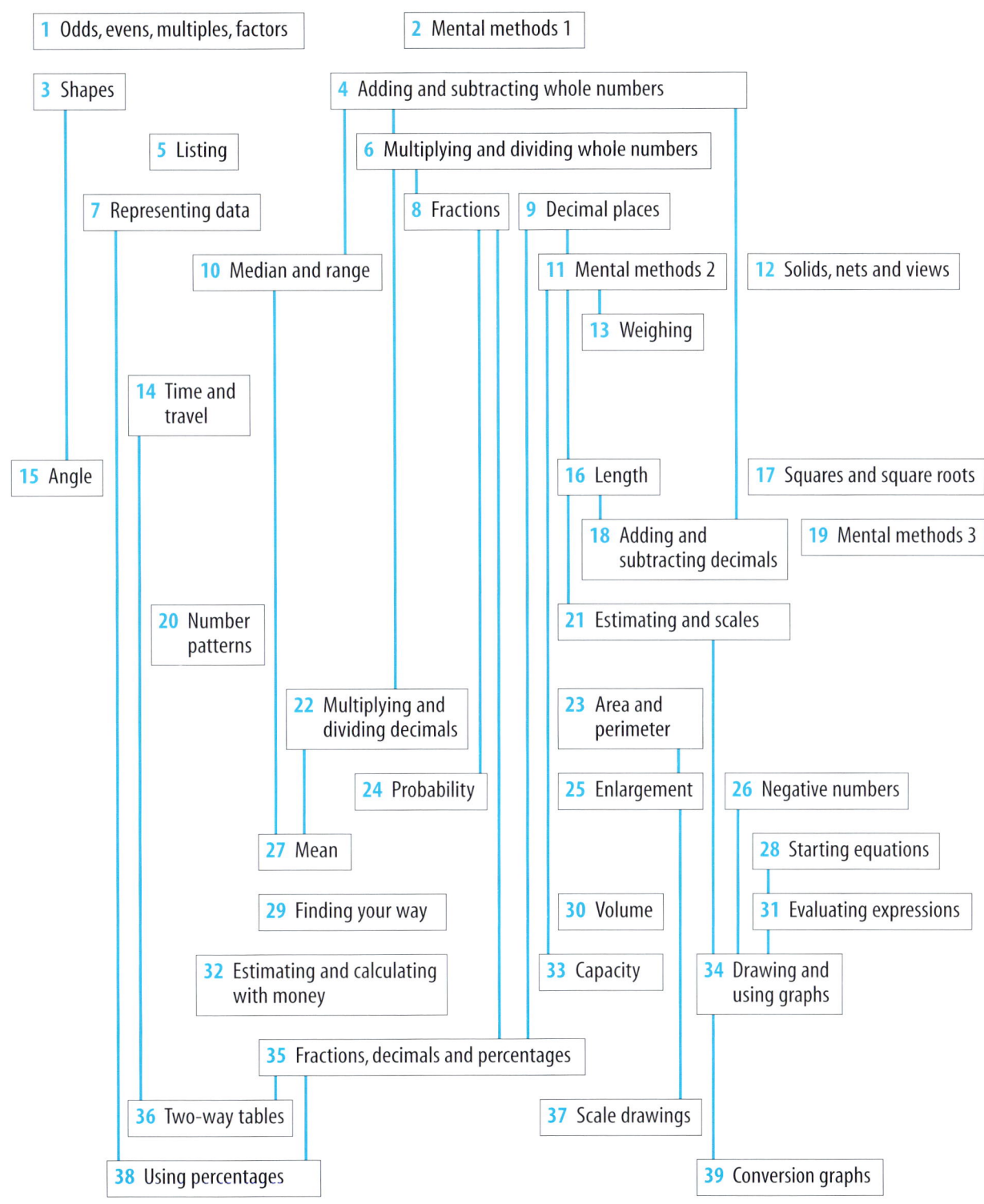

1 Odds, evens, multiples, factors

This work will help you decide when numbers are
- even or odd
- divisible by 5 or 10
- divisible by any number below 10
- multiples of numbers
- factors of numbers

A Odd and even numbers level 3

Even numbers are those that can be divided by 2 leaving no remainder.
We say that even numbers are **divisible by 2**.

Odd numbers are those that are not divisible by 2.

A1 (a) Which numbers in the loop are even?
(b) Which are odd?

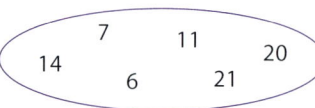

A2 Rosie runs a stall at a fete where you pick a ball out of a bag.
If you pick a ball with a number on it you win a soft toy.

- An **even** number on a ball wins a bear.

- An **odd** number on a ball wins a bunny.

What do you win with each of these, a bear or a bunny?

(a)   (b)   (c)   (d)   (e)

A3 Each of these cards shows a letter and a number.

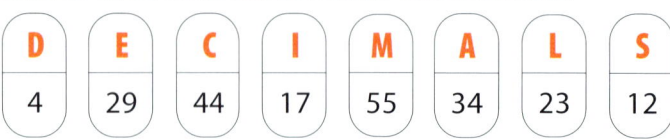

Write down each letter that is on a card with an odd number.
Rearrange these letters to give the name of a fruit.

*A4 Is the number 629 an odd or an even number?

B Divisibility
level 3

A number that can be divided by 10 exactly (leaving no remainder) is **divisible** by 10.

- Which numbers in the loop are divisible by 10?
  How can you decide quickly if a number is divisible by 10?
- Which numbers in the loop are divisible by 5?
  How can you decide quickly if a number is divisible by 5?

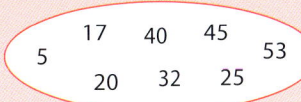

B1 (a) Which numbers in the loop are divisible by 5?

(b) Which are even?

(c) (i) Which are even **and** divisible by 5?

(ii) What do you notice about these numbers?

B2 Josie runs a stall where you pick a straw out of a box.
Each straw contains a rolled-up slip of paper with a number on it.

- A number that can be divided by 10 exactly wins £1.
- An odd number wins a balloon.
- Any other number wins a lollipop.

What do you win with each of these?

(a) 23   (b) 60   (c) 15   (d) 100   (e) 18

B3 (a) (i) Work out 21 ÷ 3.

(ii) Explain how your answer shows that 21 is divisible by 3.

(b) Which numbers in the loop are divisible by 3?

(c) Which numbers in the loop are divisible by 4?

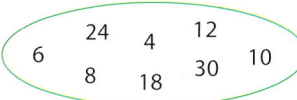

B4 Each of these cards shows a letter and a number.

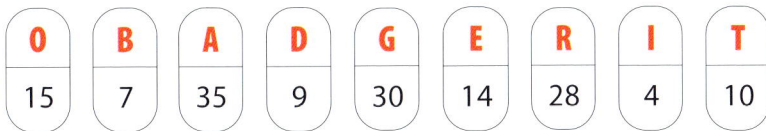

(a) Write down each letter that is on a card with

(i) a number that is divisible by 3

(ii) an even number

(iii) a number that is divisible by 5

(iv) a number that is divisible by 7

(b) Rearrange each set of letters to give the name of an animal.

1 Odds, evens, multiples, factors 9

B5 **(a)** Which numbers in the wall are divisible by 5?

(b) Which are odd?

(c) Which number is odd **and** divisible by 5?

(d) Find two numbers in the wall that add to make 230.

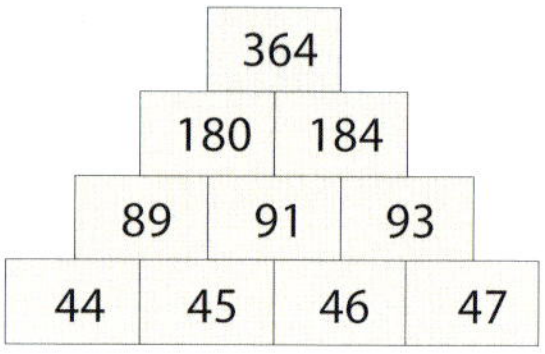

364
180 184
89 91 93
44 45 46 47

C Multiples

level 4

Multiples of 6 are numbers that are in the 6 times table: 6, 12, 18, 24, 30, …

All numbers that are divisible by 6 are multiples of 6.
For example, 60 ÷ 6 = 10 so 60 is a multiple of 6.

C1 Copy and complete this list to show all the multiples of 5 up to 50.

5, 10, 15, 20, ■, ■, ■, ■, ■, 50

C2 List all the multiples of 4 that are less than 30.

C3 Which numbers in this list are multiples of 10?

50, 72, 45, 20, 130, 68, 402, 240

C4 Which numbers in the loop are multiples of

(a) 3 **(b)** 5 **(c)** 6

15
50
16
12
39
65
36
30

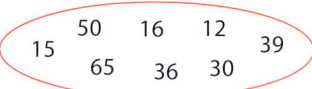

C5 Each of these cards shows a letter and a number.

S	O	N	I	C	B	R	A	W	L
77	60	55	85	27	35	15	11	132	32

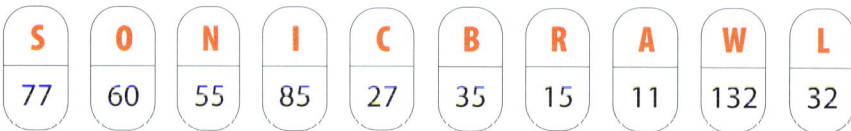

(a) Write down each letter that is on a card with

(i) a multiple of 5 **(ii)** a multiple of 4

(iii) a multiple of 3 **(iv)** a multiple of 11

(b) Rearrange each set of letters to give the name of a bird.

C6 Use the clues to find the numbers.

(a) Between 10 and 20
Multiple of 7

(b) Less than 10
Multiple of 9

(c) Between 20 and 30
Multiple of 8

(d) Between 40 and 50
Odd multiple of 5

(e) Multiple of 5 less than 60
Digits add to 8

D Factors level 4

15 can be divided exactly by 3 (15 ÷ 3 = 5).

We say that 3 is a **factor** of 15.

- Find all four factors of 15.

D1 Which number in this list is not a factor of 10?

    2, 5, 1, 3, 10

D2 Which of these numbers are factors of 12?

    2, 5, 1, 3, 10, 4, 8, 6

D3 List all the factors of 6.

D4 Which of these numbers are factors of 20?

    2, 5, 1, 3, 10, 4, 8, 6

D5 Use the clues to find the numbers.

(a) Between 2 and 8 / Factor of 9

(b) Between 5 and 10 / Factor of 28

(c) Factor of 8 / Odd number

Test yourself

T1 3 6 50 57 65 71

Copy and complete these sentences using numbers from the box. You can use a number more than once.

(a) …………………………… are all the odd numbers.

(b) …………………………… can be divided by five exactly.

(c) …………………………… can be divided by ten exactly.

(d) ………… and ………… add up to 60.

OCR

T2 Which numbers in the loop are

(a) even
(b) multiples of 10
(c) multiples of 3
(d) factors of 18

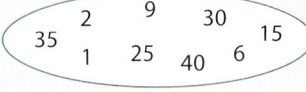

T3 Here is a list of numbers.

    32 33 36 41 45 47 50

Use a number from this list to complete each sentence.

(a) ………… is an odd number less than 40

(b) ………… is even **and** divisible by 5.

OCR

2 Mental methods 1

This work will help you
- order whole numbers
- round whole numbers to the nearest ten, hundred, thousand, …

A Place value   level 4

- Say each of these numbers.
- What does the 5 stand for in each one?

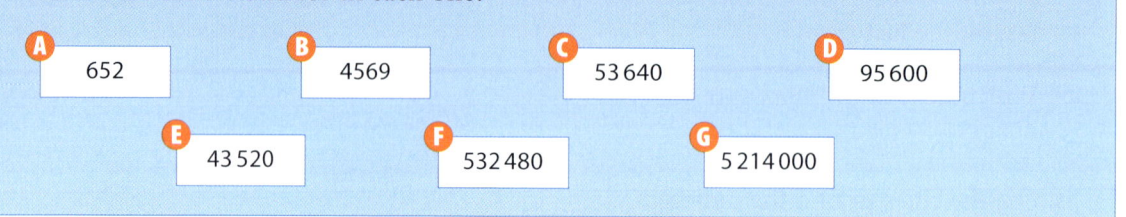

  A 652   **B** 4569   **C** 53 640   **D** 95 600

  E 43 520   **F** 532 480   **G** 5 214 000

- Match the words with the correct number.
- How would you say the unmatched number?

  3002   30 200   30 020   320   3 200 000   32 000

  three million two hundred thousand
  thirty thousand two hundred
  three thousand and two
  thirty thousand and twenty
  three hundred and twenty

A1 In the number 23 654, the figure 6 stands for 6 **hundreds** or 600.
The figure 2 stands for 2 **ten thousands**, or 20 000.

23 654

(a) What does the figure 3 stand for?
(b) What does the figure 5 stand for?

A2 In the number 497 536,

497 536

(a) what does the figure 4 stand for?
(b) what does the figure 9 stand for?

A3 Work these out in your head.
- (a) 3518 + 100
- (b) 53 763 – 1000
- (c) 42 071 + 300
- (d) 146 510 + 2000
- (e) 90 384 – 100
- (f) 471 593 + 2000
- (g) 90 342 + 5000
- (h) 276 540 – 2000

A4 Before Jan went on holiday, her car's mileometer said 57 624.
When she got back, it said 59 624.
How many miles did Jan travel on holiday?

A5 St Paul's Cathedral in London was built in 1711.
- (a) In which year was it 200 years old?
- (b) In which year will it be 2000 years old?
- (c) In which year will it be 20 000 years old (if it lasts that long!)?

A6 (a) (i) Use these cards to make the largest number you can.
    (ii) Write the number in words.
(b) (i) Use these cards to make the smallest number you can.
    (ii) Write the number in words.

A7 The table shows the total number of runs scored in test matches by some of England's best cricketers.
- (a) Which cricketer scored the most runs?
- (b) Which cricketer scored the fewest runs?
- (c) Write the numbers in order, starting with the **highest**.

Name	Total test runs
Mike Atherton	7728
Geoff Boycott	8114
Graham Gooch	8900
David Gower	8231
Alec Stewart	8463

A8 Write these lists of numbers in order, **smallest** first.
- (a) 5010   1050   1005   5100   1500
- (b) 67 342   67 531   68 752   68 200   67 275
- (c) 41 000   40 250   24 030   20 840   38 040

B Rounding   level 4

Newspapers usually round the numbers they use in headlines.
How would you round the numbers in these headlines?

657 people injured

Damages of £316 254 awarded to peer

23 457 people attend concert

3528 jobs created

2 Mental methods 1   13

B1 Round these numbers to the nearest ten.
(a) 83   (b) 25   (c) 436   (d) 912   (e) 3288

B2 Round these numbers to the nearest hundred.
(a) 523   (b) 396   (c) 2618   (d) 5784   (e) 32 431

B3 Round these numbers to the nearest thousand.
(a) 7456   (b) 8097   (c) 2518   (d) 18 309   (e) 7956

B4 The world's longest river is the Nile. It is 4132 miles long.
Round the number 4132 to the nearest
(a) hundred   (b) ten   (c) thousand

B5 The distance from London to Wellington (New Zealand) is 11 682 miles.
Round the number 11 682 to the nearest
(a) thousand   (b) ten thousand   (c) hundred   (d) ten

B6 Round
(a) 46 023 to the nearest thousand
(b) 68 086 to the nearest hundred
(c) 647 254 to the nearest ten thousand
(d) 7 835 091 to the nearest million

B7 The table shows the greatest depths, in metres, of the world's oceans.
(a) Copy the table and write the depths to the nearest thousand metres.
(b) Which ocean is closest to 10 000 m deep?

Ocean	Greatest depth
Pacific Ocean	10 924 m
Atlantic Ocean	9 219 m
Indian Ocean	7 455 m
Arctic Ocean	5 625 m

Test yourself

T1 (a) Write the number seventeen thousand, two hundred and fifty-two in figures.
(b) Write the number 5367 correct to the nearest hundred.
(c) Write down the value of the 4 in the number 274 863.
Edexcel

T2 Write the number 2764
(a) to the nearest 10   (b) to the nearest 100
AQA

T3 (a) Write in figures six thousand five hundred and eighty.
(b) (i) Write 4873 in words   (ii) Write 4873 to the nearest hundred.
(c) Arrange these numbers in order of size, starting with the smallest.
7510   7150   5107   7105
OCR

14   2 Mental methods 1

3 Shapes

This work will help you

- use the names of the different parts of a circle
- name types of triangles, quadrilaterals and polygons
- plot coordinates in the first quadrant
- recognise the lines of symmetry of a shape
- complete shapes given a vertical or horizontal line of symmetry

You need a pair of compasses and sheet FT–1.

A Circles level 3

This diagram shows some different parts of a circle.

A **radius** is a straight line from the centre to the edge.

A **diameter** is a straight line through the centre from edge to edge.

The **circumference** is the curved edge of a circle.

A1 (a) Which two of these lines are diameters of the circle?
(b) Measure the diameter in centimetres.
(c) What is the length of the radius of this circle?

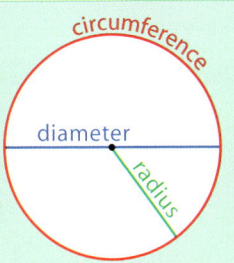

A2 This is one of the world's largest yo-yos. The boy is 120 cm tall.

Roughly, what is the length of the radius of the yo-yo?

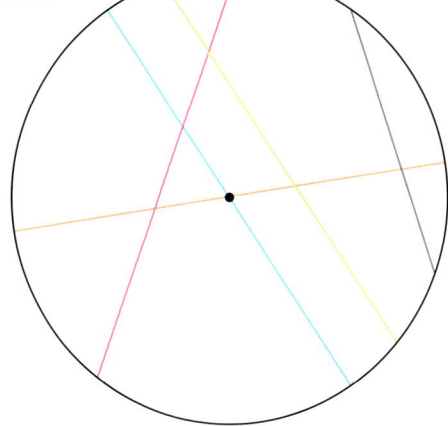

3 Shapes 15

A3 (a) Which of the marked points is the centre of
  (i) the smaller circle
  (ii) the larger circle
(b) Measure the length of
  (i) the diameter of the larger circle
  (ii) the radius of the smaller circle

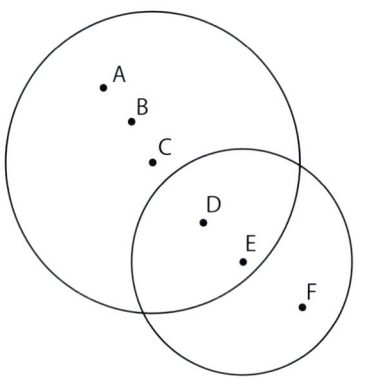

A4 Use a pair of compasses to draw a circle that has
  (a) a radius of length 4 cm   **(b)** a diameter of length 10 cm

B Triangles
level 4

A **scalene** triangle has edges that are all different lengths.
These triangles are all scalene.

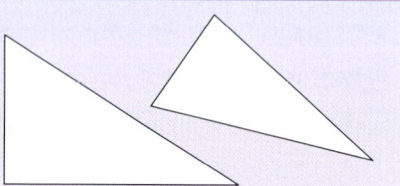

B1 Which of these are scalene triangles?

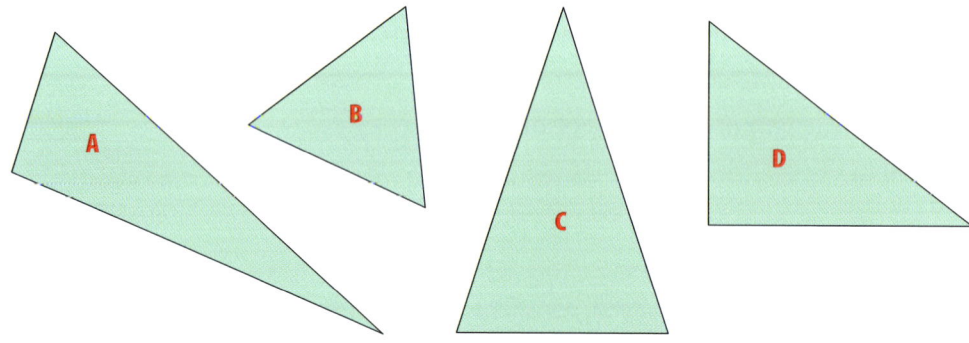

An **isosceles** triangle has two edges that are the same length.
These triangles are all isosceles.

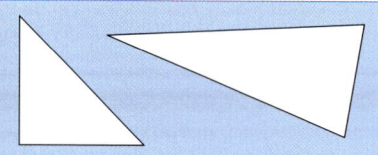

B2 Measure the edges of this triangle. Is it an isosceles triangle?

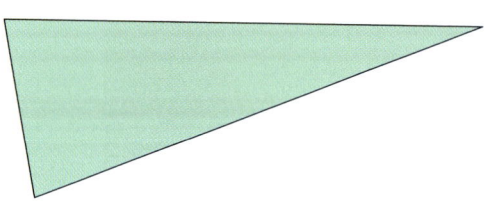

16 3 Shapes

B3 Which of these are isosceles triangles?

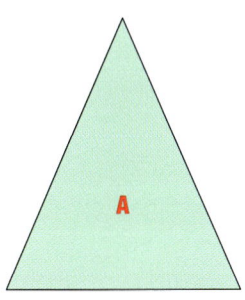
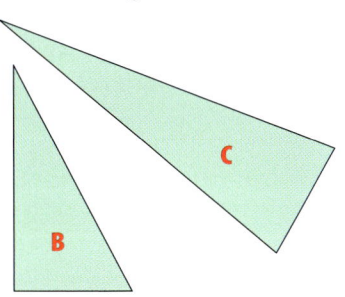
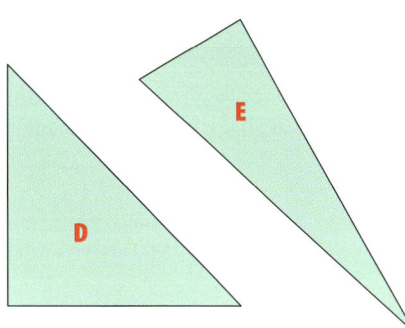

An **equilateral** triangle has edges that are all the same length.

These triangles are all equilateral.

B4 Which of these are equilateral triangles?

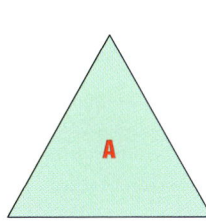
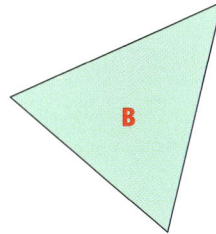
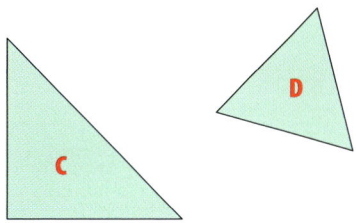

B5 Draw a coordinate grid like this one that goes from 0 to 10 in both directions.

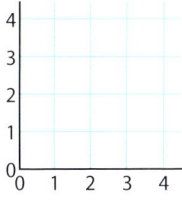

(a) (i) Plot the points given by the coordinates (2, 1), (1, 5) and (3, 7) and join them up with straight lines.

(ii) What kind of triangle have you drawn?

(b) (i) Plot the points (5, 4), (10, 9) and (3, 10) and join them up.

(ii) What kind of triangle have you drawn?

(c) Plot the points (8, 3), and (10, 3) on your grid.
Give the coordinates of another point so that the three points join to make an isosceles triangle.

B6 Copy this triangle on to squared paper.

(a) What kind of triangle is it?

(b) The triangle has one line of symmetry.
Show the line of symmetry clearly on your diagram.

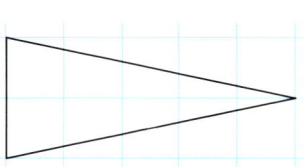

3 Shapes 17

C Quadrilaterals
level 4

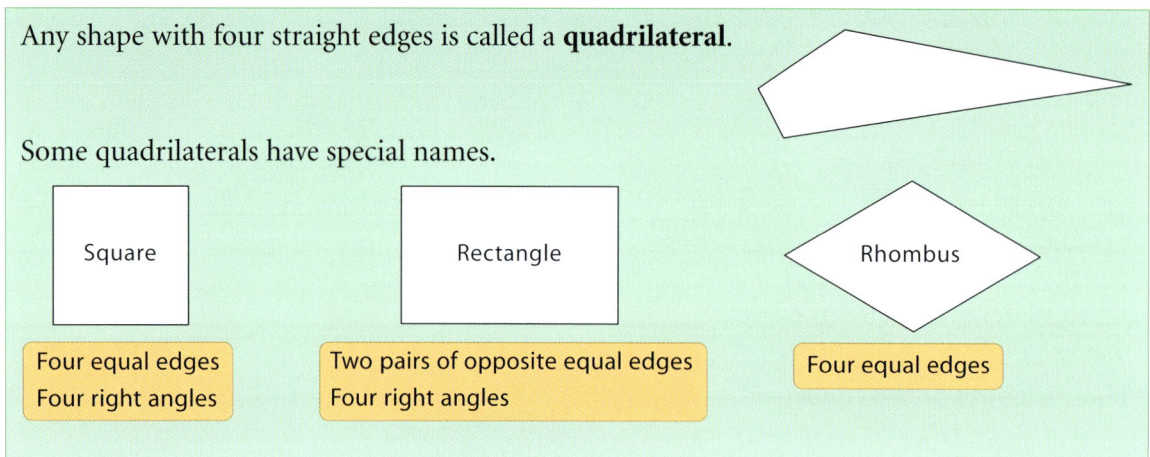

Any shape with four straight edges is called a **quadrilateral**.

Some quadrilaterals have special names.

Square	Rectangle	Rhombus
Four equal edges Four right angles	Two pairs of opposite equal edges Four right angles	Four equal edges

C1 (a) Which of these quadrilaterals are squares?

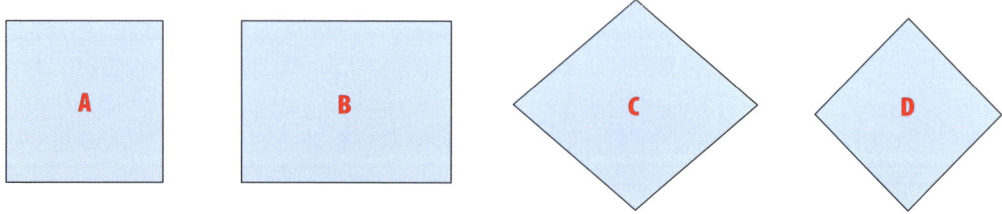

  (b) Give one rectangle from this set of shapes.

  (c) Give one rhombus from this set of shapes.

C2 Draw a coordinate grid like this that goes from 0 to 10 in both directions.

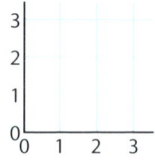

  (a) (i) Plot the points (1, 1), (1, 4) and (4, 4) on your grid.
     Plot another point so that the four points are the corners of a square.
  (ii) Join the points up and draw all the lines of symmetry on your square.

  (b) Plot the points (7, 6), (5, 3) and (7, 0) on your grid.
     Plot another point so that the four points are the corners of a rhombus.
     Write down the coordinates of this fourth point.
     Join the points up.

  (c) Mark the points (0, 7), (2, 5) and (5, 8) on your grid.
     Mark another point so that the four points are the corners of a rectangle.
     Label each point with its coordinates.

18 3 Shapes

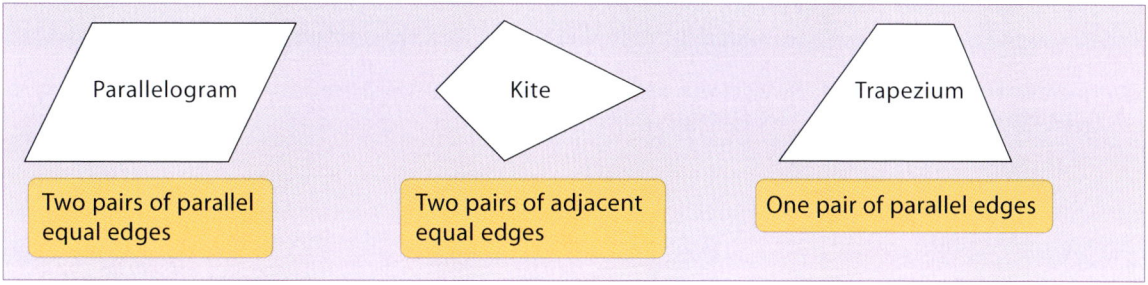

C3 Write down all of the quadrilaterals below that are

(a) kites (b) trapeziums

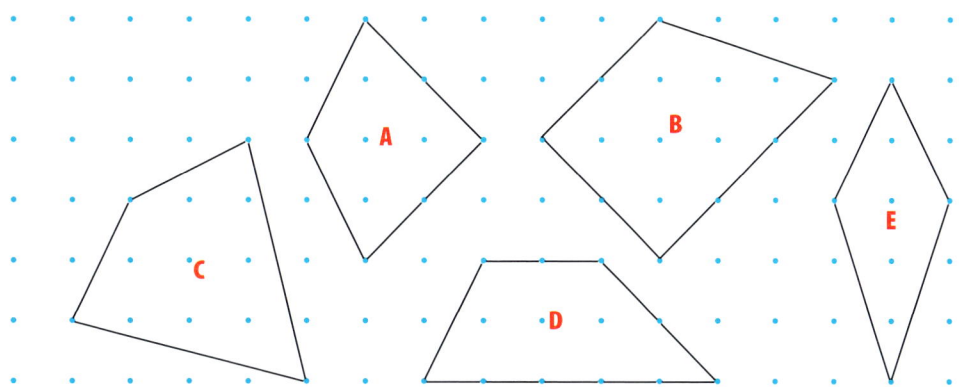

C4 Copy the diagram on the right.

(a) Mark a fourth point on your diagram so that the four points join to make a parallelogram.

(b) What are the coordinates of the fourth point?

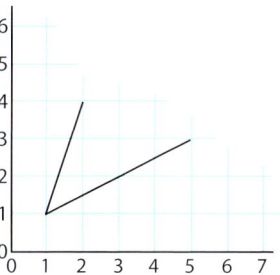

C5 For each diagram below

(i) copy and complete it so that the dotted line is a line of symmetry

(ii) write down the name of the compete shape

(a) 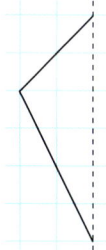 (b)  (c)

3 Shapes 19

D More than four edges   level 4

A **pentagon** is a shape with five straight edges.
Both these shapes are pentagons.

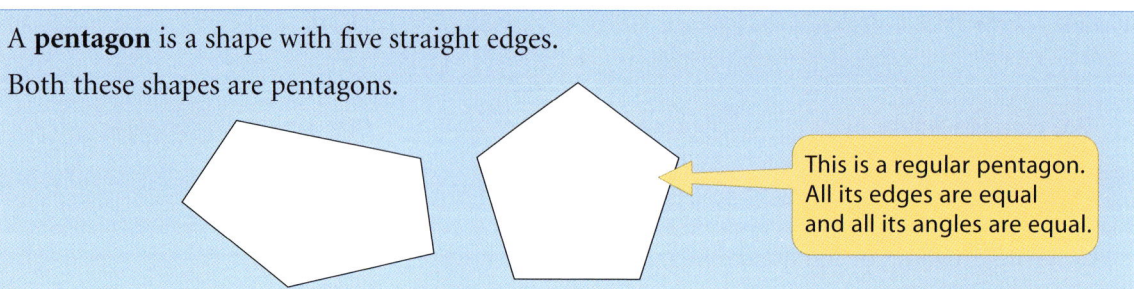

This is a regular pentagon. All its edges are equal and all its angles are equal.

D1 (a) In this tiling pattern, how many of the tiles are pentagons?
     (b) What shape are the orange tiles?

A **hexagon** is a shape with six straight edges.
Both these shapes are hexagons.

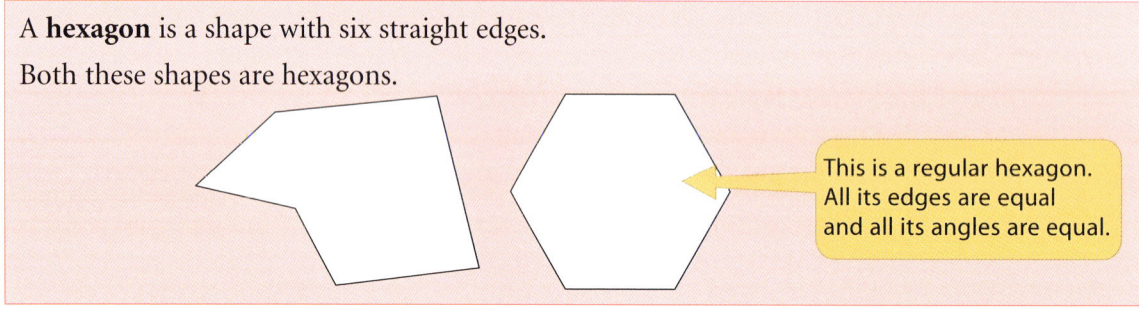

This is a regular hexagon. All its edges are equal and all its angles are equal.

D2 (a) In this tiling pattern, how many of the tiles are hexagons?
     (b) State the colour of the rhombus-shaped tiles.
     (c) What shape are the grey tiles?

20 3 Shapes

An **octagon** is a shape with eight straight edges.
Both these shapes are octagons.

This is a regular octagon. All its edges are equal and all its angles are equal.

D3 (a) From this set of shapes, list
  (i) all the octagons
  (ii) all the pentagons

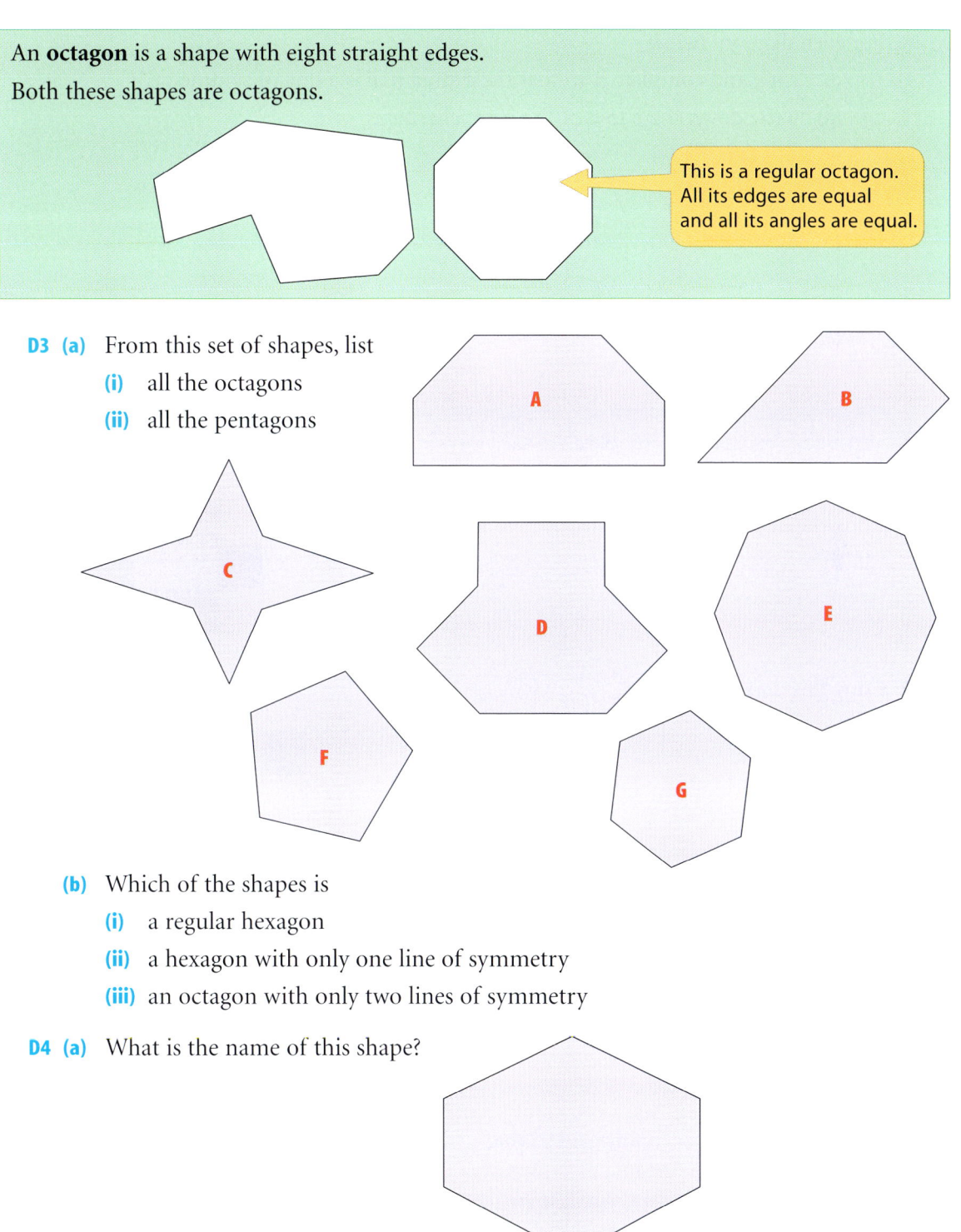

(b) Which of the shapes is
  (i) a regular hexagon
  (ii) a hexagon with only one line of symmetry
  (iii) an octagon with only two lines of symmetry

D4 (a) What is the name of this shape?

(b) How many lines of symmetry does it have?

D5 (a) Draw a pentagon that has only one line of symmetry.
(b) Show the line of symmetry on your shape.

D6 For each diagram below
    (i) copy and complete it so that the dotted line is a line of symmetry
    (ii) write down the name of the compete shape

(a) 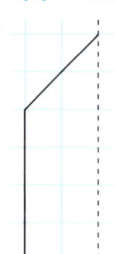   **(b)** 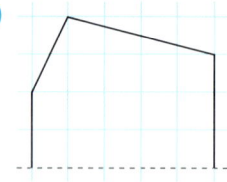   **(c)**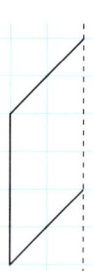

E Shading squares to give reflection symmetry

E1 This pattern has two lines of symmetry.
Copy the diagram and draw the lines of symmetry on it.

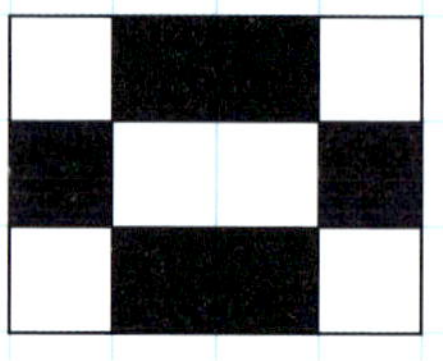

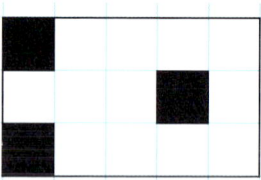

E2 Copy this diagram.
Shade in three more squares so that the diagram has two lines of symmetry.

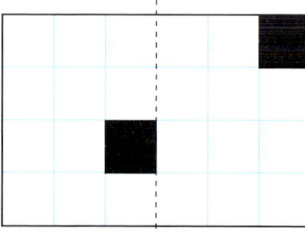

E3 Copy this diagram.
Shade in six more squares so that the dotted line is the **only** line of symmetry.

E4 Sketch all the different ways you can find of shading three squares in this grid so that the pattern has only one line of symmetry.
Draw the line of symmetry on each diagram.

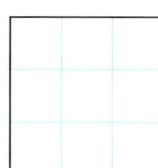

22 3 Shapes

Test yourself

T1 This wheel is the London Eye.
The top of the wheel is 135 m above the ground.
Estimate the length of the radius of the wheel.

T2 Draw a coordinate grid that goes from 0 to 6 in both directions.
  (a) Plot the points given by the coordinates (1, 1), (3, 6) and (5, 1) and join them up.
  (b) What kind of triangle have you drawn?

T3 (a) What is the name of this shape?

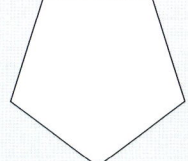

  (b) How many sides does a hexagon have?

T4 Draw a coordinate grid that goes from 0 to 8 in both directions.
  (a) Mark the points given by the coordinates (0, 4), (2, 1) and (6, 4).
  (b) (i) Mark another point so that the four points are the corners of a kite.
       Join the points up.
     (ii) Label each point with its coordinates.

T5 Copy and complete this diagram so that the dotted line is a line of symmetry.

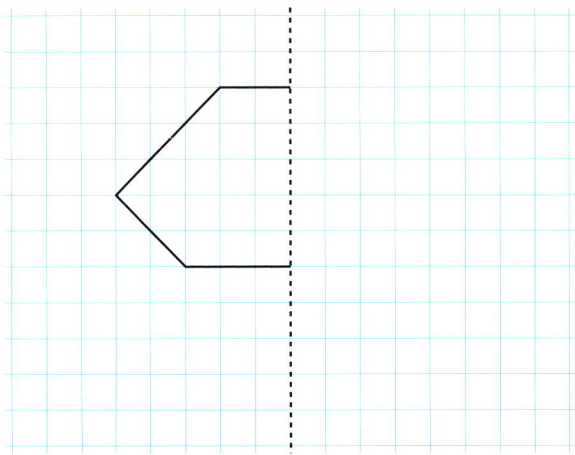

OCR

T6–T8 These questions are on sheet FT–1.

4 Adding and subtracting whole numbers

This work will help you add and subtract whole numbers without using a calculator.
You need sheet FT–2.

A Using written and mental methods   level 3

- Can you fit these digits 3 4 4 8 into this grid
  to make the biggest possible result?
  What is the smallest possible result with these digits?
- Try this with some different sets of digits.

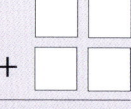

- Can you fit these digits 1 8 6 9 into this grid
  to make the biggest possible result?
  What is the smallest possible result with these digits?
- Try this with some different sets of digits.

A1 In a magic square the numbers in each row, each column and each diagonal add to the same total.

11	6	7
4	8	12
9		5

(a) What is the total for this magic square?

(b) What is the missing number in the bottom row?

A2 Copy and complete these magic squares.

(a)
22	2	15
	13	

(b)
15	19	23
20		

(c)
30		36
	38	
		46

(d)
		82
		74
90		102

A3 Work these out.

(a) 16 + 41   (b) 29 + 53   (c) 68 + 45

(d) 164 + 528   (e) 576 + 95   (f) 36 + 207 + 82

A4 Work these out.

(a) 58 – 32   (b) 81 – 73   (c) 74 – 27

(d) 178 – 53   (e) 463 – 147   (f) 323 – 158

A5 Do the puzzles on sheet FT–2.

> The **difference** between two numbers can be found by taking the smaller from the larger.
> For example, the difference between 24 and 39 is the result of 39 – 24 which is 15.

A6 Some young farmers run a competition to guess the weight of a pig.
Five guesses are shown.

A 391 kg   **B** 329 kg   **C** 383 kg   **D** 325 kg   **E** 387 kg

The real weight of the pig is 358 kg.

(a) What is the difference between each guess and the real weight?

(b) Which guess is closest?

A7 (a) A primary school has 91 girls and 93 boys.
How many pupils go to the school altogether?

(b) One day 115 of these pupils brought a packed lunch.
How many pupils did not bring a packed lunch?

A8 Copy and complete these calculations, filling in the missing digits.

(a) ▪1
   + 2▪
   ─────
     8 8

(b) 5▪
   + ▪4
   ─────
     9 1

(c) ▪7
   + 3▪
   ─────
   ▪2 6

(d) 7 6
   − 4▪
   ─────
     2 8

(e) 4▪
   − ▪9
   ─────
     2 3

> The **sum** of a set of numbers is the result of adding them together.

A9 (a) Find two numbers in the loop with a sum of
   (i) 59   (ii) 34
   (iii) 127   (iv) 103

(b) Find two numbers in the loop with a difference of
   (i) 13   (ii) 9

   (25 56 71
    14   9   35
     12 47   68)

Test yourself

T1 Work these out.

(a) 23 + 48   (b) 37 + 85   (c) 64 − 26   (d) 105 − 47

T2 The Angel Falls and the Tugela Falls are the world's highest waterfalls.
The height of the Angel Falls is 979 metres.
The height of the Tugela Falls is 853 metres.
What is the difference between their heights?

T3 Find the sum of 34, 13 and 161.

5 Listing

This work will help you list all the outcomes in a given situation.

You need sheets FT–3 and FT–4.

A Arrangements   level 4

If you have a red block and a blue block,   there are only two different ways to build a tower.

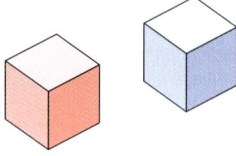   

If you have three blocks,   here are two different towers.

These ways are shown in this table.

Top block	Green	Blue
Middle block	Blue	Red
Bottom block	Red	Green

There are more ways to build a tower using these three blocks.

- How many different towers of three blocks can you find?
  Show all of your ways in a table.
- Check that you haven't included the same tower twice.
- Check that you haven't missed out any possible towers.

When there are lots of ways to arrange a group of objects it is important to be **systematic**.
This makes it easier to check that you have included everything and have not repeated any arrangements.

A1 You have these three cards.
Find all the different ways you can arrange
the three cards in a row.

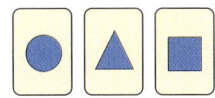

Show all the arrangements in the table on sheet FT–3.

A2 Anita has these three letter cards.
She wants to find all of the different arrangements
of the three cards in a row.

(a) Copy and complete this table to show all
of the ways she can arrange the three cards.

(b) How many of the arrangements make a word?

First	Second	Third
A	E	P

A3 Martin has these three number cards.

(a) Copy and complete this table to show all
the different ways he can arrange the cards.

(b) What is the largest number he can make?

(c) What is the smallest number he can make?

First	Second	Third
8	3	5

Challenge

How many different towers can you make
using these four blocks?

Make a table to show your arrangements.
Make sure your list is systematic.

B Combined choices   level 4

Five students from class 10T want to be on the school council.

The class needs to choose one boy and one girl.

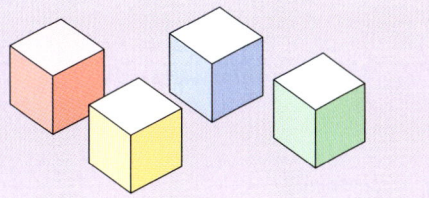

- How many different pairs of students are there?
  Show all of the pairs in a table.

Boy	Girl
Daniel	Emily
Daniel	Mia

- How many different pairs of students would there be if Kyle
  also wanted to be on the school council?

5 Listing 27

B1 Martha's tea shop offers a morning special.
You can choose a drink and a snack.

Find all the combinations of drink and snack.
Use the table on sheet FT–4.

Drinks
Tea
Coffee
Hot Chocolate

Snacks
Muffin
Cookie
Bagel

B2 Joe and Naz go to a sports camp.
Each person can choose one activity from these three.

List all the possible combinations of their choices.
The list has been started for you.

Football   Athletics   Cricket

Joe	Naz
Football	Football
Football	Athletics

B3 Jasmine has English, Maths and History homework.

(a) List all the ways she can choose to do her homework.
The list has been started for you.
E stands for English, M for Maths and H for History.

First	Second	Third
E	M	H

(b) Jasmine always does her English homework last.
How many ways are there with English last?

Test yourself

T1 Troy has these three cards.

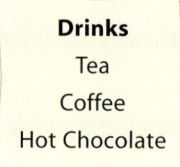

He can arrange them in different ways to make 3-digit numbers.

(a) List all the possible 3-digit numbers he can make.
Copy and complete the table.
One has been done for you.

5	4	6

You may not need all the lines.

(b) How many of the numbers in the table are greater than 500?
(c) Which is the largest number in the table?

OCR

6 Multiplying and dividing whole numbers

This work will help you
- multiply a two-digit number by a one-digit number
- divide a two-digit number by a one-digit number and find remainders

A Multiplying whole numbers   level 4

```
  6 4                              60   4
× 3          ← 64 × 3 →        3 │ 180  12
─────
1 9 2                            180 + 12 = 192
  1
```

A1 Work these out.
(a) 14 × 6   (b) 13 × 9   (c) 27 × 4   (d) 32 × 5
(e) 46 × 3   (f) 31 × 8   (g) 74 × 5   (h) 39 × 7

A2 These three cards can be arranged to make multiplications.

(a) Copy and complete this multiplication.

3 5 × 2 =

(b) Use the cards to make a multiplication with result
(i) 75   (ii) 115   (iii) 106

(c) What is the largest result you can make with these cards?

A3 It costs Sam 5p to send a text message.
How much does it cost him to send 28 text messages?

A4 Some small bottles of beer are packed in boxes of 24.
How many bottles are there altogether in 7 boxes?

A5 A coach holds 52 passengers.
How many passengers are there in 6 full coaches?

A6 There are 16 teams taking part in a netball competition.
Each team has 7 players.
How many players are there altogether in the competition?

B Dividing whole numbers level 4

51 ÷ 3 $\begin{array}{r}1\,7\\3\,)\overline{5\,^{2}1}\end{array}$ 97 ÷ 5 $\begin{array}{r}1\,9\text{ remainder 2}\\5\,)\overline{9\,^{4}7}\end{array}$

B1 Work these out.
- (a) 38 ÷ 2
- (b) 42 ÷ 3
- (c) 68 ÷ 4
- (d) 85 ÷ 5
- (e) 76 ÷ 4
- (f) 90 ÷ 6
- (g) 91 ÷ 7
- (h) 96 ÷ 8

B2 Yoghurts are sold in packs of 4.
How many packs can be filled from 56 yoghurts?

B3 Chocolate biscuits are sold in packs of 7.
How many packs can be filled from 84 biscuits?

B4 Work these out. Each answer has a remainder.
- (a) 34 ÷ 3
- (b) 57 ÷ 2
- (c) 73 ÷ 5
- (d) 49 ÷ 4
- (e) 70 ÷ 6
- (f) 96 ÷ 7
- (g) 93 ÷ 9
- (h) 86 ÷ 3

B5 Dean is making Easter baskets.
He has 46 chocolate eggs.
How many baskets can he fill if he puts 3 eggs in each?

B6 Each table in a café seats 4 people.
How many tables will be needed for 83 people?

B7 Cakes are sold in packs of 6.
How many packs can be filled from 80 cakes?

Test yourself

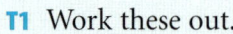

T1 Work these out.
- (a) 47 × 6
- (b) 84 ÷ 3
- (c) 51 × 8
- (d) 92 ÷ 4

T2 Apples are sold in bags of 8.
- (a) How many apples are there in 27 bags?
- (b) How many bags can be filled from 96 apples?

T3 There are 50 people waiting for a ride at a theme park.
Each car can carry 4 people.
How many cars are needed?

AQA

Review 1

1 (a) Find two numbers between 20 and 30 that are divisible by 7.
  (b) List all the factors of 8.

2 The distance from Bristol to Inverness is 867 km.
  Round this to the nearest 100 km.

3 How many sides has an octagon?

4 (a) Use these cards to make as many numbers as you can that are larger than 100.
  (b) Which of your numbers are divisible by 10?
  (c) (i) Write down the largest of your numbers.
    (ii) What does the 4 stand for in this number?

5 (a) The mileometer on Joan's new car shows eight thousand and thirty miles.
    Write this distance in figures.
  (b) She travels 100 miles.
    (i) What does the mileometer show now?
    (ii) Write this distance correct to the nearest thousand miles.

6 Work these out.
  (a) 207 + 34   **(b)** 36 × 5   **(c)** 762 − 84   **(d)** 87 ÷ 3

7 Copy this diagram.
  Shade three more squares so that the pattern has just two lines of symmetry.
  Show the lines of symmetry.

8 Write this list of numbers in order, smallest first.
   6400   6004   4060   4006   4600

9 It costs Kelly 12p to send a text message.
  How much does it cost her to send 8 text messages?

10 Draw a coordinate grid that goes from 0 to 8 in both directions.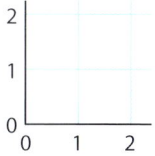
  (a) Mark the points given by the coordinates (3, 0), (6, 4) and (3, 8).
  (b) Mark another point so that the four points are the corners of a rhombus. Join the points up.
  (c) Label each point with its coordinates.
  (d) How many lines of symmetry has the rhombus?

Review 1 31

7 Representing data

This work will help you

- show a frequency distribution as a table, chart or pictogram
- find the mode

A Frequency charts and mode (for types of things)   level 4

Chris is working on a traffic survey.
His job is to record the colour of each car that goes past him.

He uses a **tally table**.

Here is his table after an hour of recording.

Colour of car	Tally	Frequency
White	IIII IIII IIII I	16
Red	IIII IIII IIII IIII II	22
Blue	IIII IIII III	13
Green	IIII	5
Other	II	2

Frequency means the number of times something happens.
For example, Chris saw 16 white cars, so the frequency of 'white' is 16.

This **frequency chart** shows Chris's data.

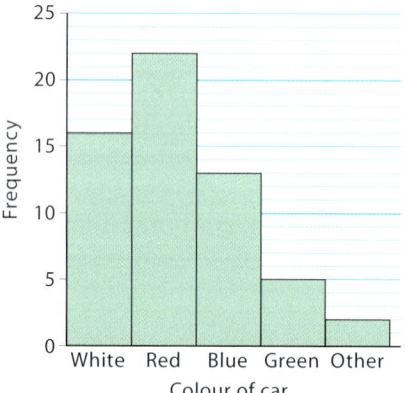

The most frequent colour is called the **mode**.
The mode is 'red'.

A1 Cara records the country where each car is made.
Here is her tally table.

(a) What is the frequency of British cars?

(b) What is the frequency of German cars?

(c) Which country is the mode?

(d) Draw a frequency chart for the data.

Country	Tally
Britain	IIII IIII II
France	IIII III
Germany	IIII IIII I
Other European	IIII II
Japan	IIII IIII III
Other	IIII III

32   7 Representing data

A2 Gemma asks the people in her class what animals they have.
Her record is shown below.

D C C G C D D H O R D C O C G H C R O D D C C G G C D R H H G C C D

C = cat, D = dog, G = guinea pig, H = hamster, R = rabbit, O = other

(a) Make a tally table for this data.

(b) Draw a frequency chart for the data.

(c) Which animal is the mode?

A3 Manoj asks people their favourite flavour of crisps.
His results are shown in this frequency chart.

(a) How many people like 'Ready salted' best?

(b) How many people like 'Chilli beef' best?

(c) Which flavour is the mode?

(d) How many people did Manoj ask altogether?

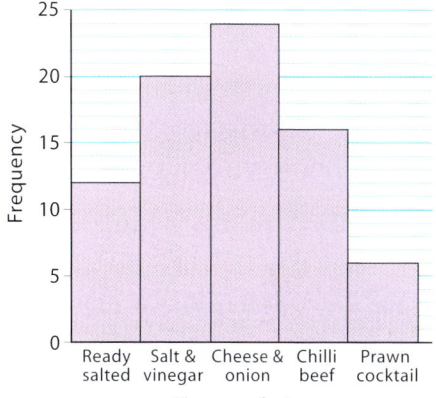

A4 A hotel records where each guest comes from.
Here is a summary of the results.

Part of the world	Frequency
Europe	36
Asia	17
Africa	10
North America	48
South America	6
Australasia	15

To draw a frequency chart, you need to decide what scale to use.
The highest frequency is 48, so make a scale that goes up to 50.

Two possible scales are shown on the right.

Choose your scale and draw the frequency chart.

A5 A café records the hot drinks people buy.
Here are the results.

Drink	Frequency
Tea	94
White coffee	66
Black coffee	13
Hot chocolate	38
Herbal tea	10

Draw a frequency chart for this data.

7 Representing data   33

B Dual bar charts

A school needs to decide on a colour for its uniform.
The choice is between black, blue, brown, green or maroon.

A survey is carried out to see whether boys and girls have different choices.
The results are shown in a **dual bar chart**.

The bars are in pairs, one for boys and one for girls.

There is a gap between each dual bar, to make the chart easier to read.

B1 (a) How many boys chose brown?
   (b) How many girls chose green?
   (c) What was the boys' most popular colour?
   (d) What was the girls' least popular colour?
   (e) If boys' and girls' votes are added together, which colour is most popular?

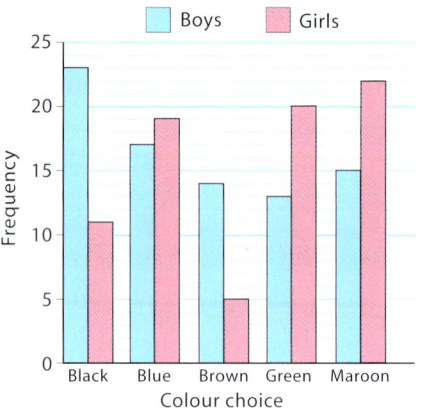

B2 This dual bar chart shows the numbers of goals scored in home and away matches by five football teams.

   (a) How many goals did City score in away matches?
   (b) Which team scored the most goals in home matches?
   (c) Which teams scored more goals in away matches than in home matches?

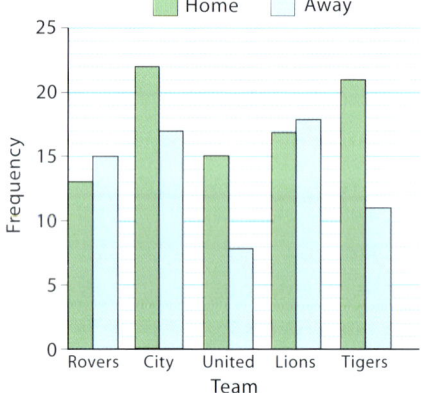

B3 A school trip is to be organised.
There are six possible places to go.

The chart shows girls' and boys' choices.

   (a) Which is the boys' least popular choice?
   (b) Which is the girls' most popular choice?
   (c) Which place gets the most votes overall?

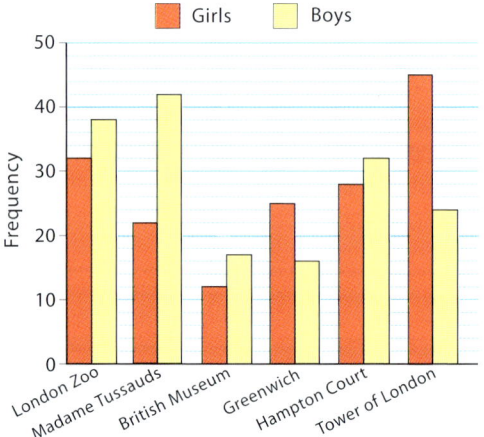

34   7 Representing data

B4 This table shows the numbers of male and female animals on a farm.

Animal	Cattle	Sheep	Pigs	Goats
Number of males	15	27	9	7
Number of females	33	30	15	4

Draw a dual bar chart to show this data.

B5 This table shows the number of boys and girls taking certain subjects at a college.

Subject	English	Maths	Sciences	History	Geography
Number of boys	22	27	12	7	11
Number of girls	30	19	10	23	8

Draw a dual bar chart to show this data.

C Pictograms   level 4

This diagram shows the numbers of trees of different kinds growing in a wood.

The diagram is called a **pictogram**.

The key at the top tells you that each symbol stands for 10 trees.

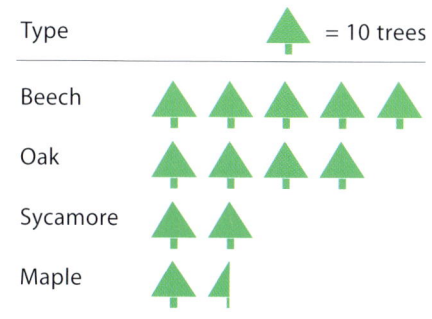

C1 (a) How many beech trees are there in the wood?

(b) What does stand for?

(c) How many maple trees are there?

(d) How many trees are there altogether?

C2 This table shows how many jars of different kinds of jam were sold in a shop last week.

Type of jam	Strawberry	Raspberry	Plum	Apricot	Blackcurrant
Number sold	80	40	20	60	30

Copy and complete this pictogram to show the data.

Type of jam   = 20 jars
Strawberry
Raspberry
(and so on)

C3 Draw a pictogram to show this data. Choose your own symbol for 10 cars.

Type of car	3-door hatchback	5-door hatchback	2-door saloon	4-door saloon	Estate
Number sold	20	40	10	35	15

7 Representing data   35

D Frequency charts and mode (for quantities) — level 4

Jack and Sarah are working on a traffic survey.
They are counting the number of people in the cars that go past them.
Jack does the cars going one way and Sarah the other way.

Jack records his numbers in a list: 3 1 2 2 1 4 3 4 1 1 2 5 3 2 4 3 1 2 1 ...

Sarah makes a tally chart for her numbers:

Number of people in car	Tally	Frequency
1												12
2									8
3							6
4					3
5				2

D1 Here is Jack's complete list for 35 cars.

3 1 2 2 1 4 3 4 1 2 2 5 3 2 4 3 1 2 1 2 2 1 3 4 2 3 1 1 2 5 3 1 1 2 3

(a) Make a tally chart for Jack's numbers and complete the frequency column.

(b) Add the frequencies. Check that you get 35, the total number of cars.

D2 This frequency chart shows Sarah's data.

(a) Which number of people is the mode?

(b) How many cars did Sarah count altogether?

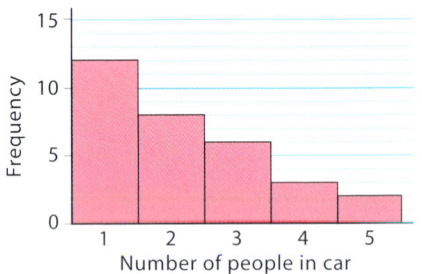

D3 (a) Draw a frequency chart for Jack's data in question D1.

(b) What is the mode?

D4 Rana counts the number of eggs in birds' nests. Here is her record.

3 2 0 4 2 4 0 0 3 5 1 5 0 0 4 2 0 3 4 0 2 1 0 4 0 3 4 4 2 5 0 0 1 2 3

(a) Make a tally chart for these numbers.

(b) Draw a frequency chart.

(c) What is the mode?

D5 Bruce counts the matches in boxes.
His tally chart is shown on the right.
Draw a frequency chart for the data.

Number of matches in a box	Tally
44							
45												
46										
47					

36 0 Representing data

D6 This list shows the numbers of goals scored in 30 football matches in a season.

2 0 3 2 0 1 1 4 2 3 2 2 2 3 3 5 0 3 1 3 1 0 4 5 3 1 4 0 1 1 4

(a) Make a tally chart for these numbers.

(b) Draw a frequency chart.

(c) What is the mode?

Test yourself

T1 The bar chart shows the number of packets of different flavours of crisps sold at a disco.

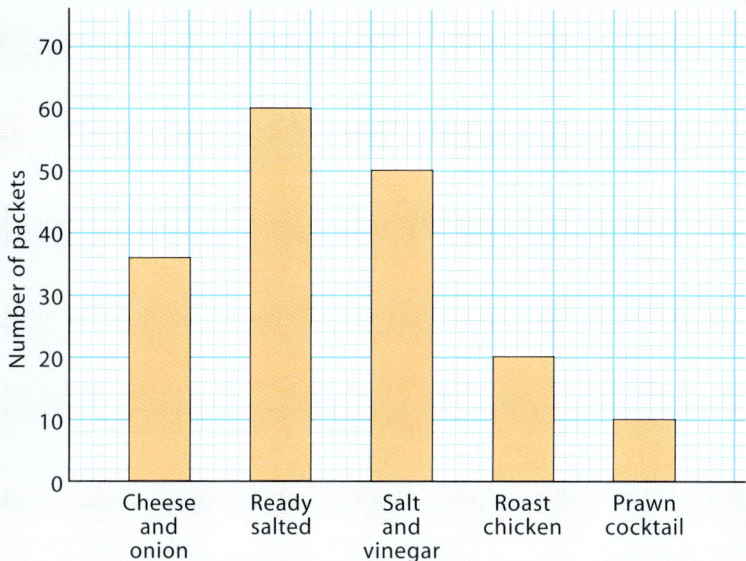

(a) Which flavour sold the most?

(b) How many packets of cheese and onion crisps were sold?

(c) How many more packets of salt and vinegar crisps than packets of roast chicken crisps were sold?

(d) How many packets of crisps were sold altogether?

AQA

T2 Nina took the sweets from a packet one by one and noted their colours. Here is her record.

R Y R R G Y O R G O O G Y Y O R R Y O O G G R Y O O R G G Y R R R G

R = red, Y = yellow, G = green, O = orange

(a) Copy and complete this tally chart.

(b) Which colour is the mode?

(c) Draw a frequency chart to show the data.

Colour	Tally	Frequency
Red		
Yellow		
Green		
Orange		

7 Representing data 37

T3 The bar chart shows the numbers of boys and girls who went to the cinema each day last week.

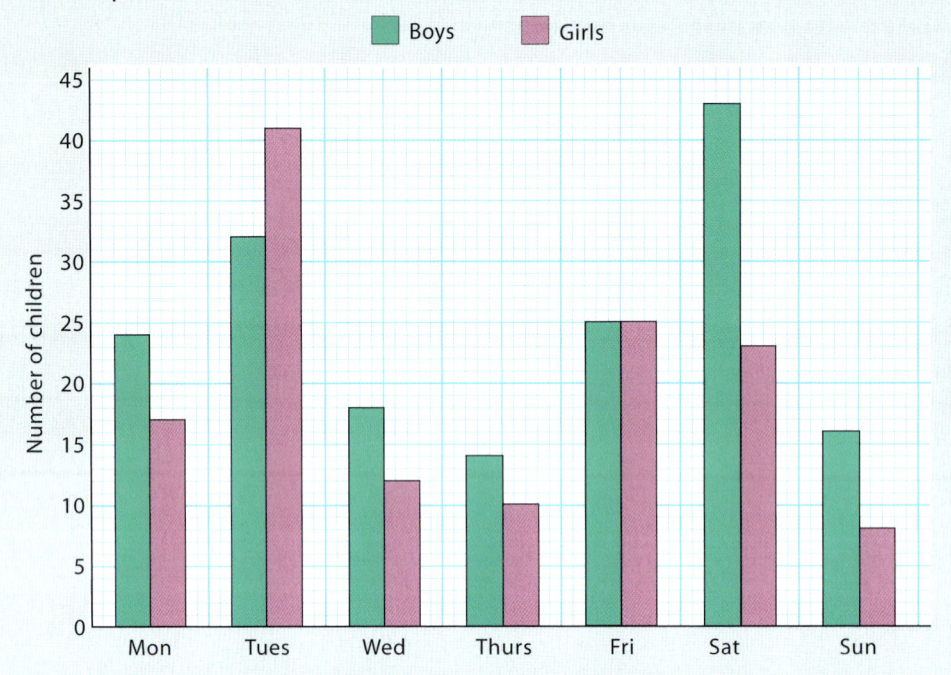

(a) How many boys went to the cinema on Monday?
(b) On which day did most girls go to the cinema?
(c) On which day was the number of boys equal to the number of girls?
(d) How many more boys than girls went to the cinema on Wednesday?
(e) Kirk says, 'On Sunday twice as many boys as girls went to the cinema.' Is he correct? Explain your answer.

AQA

T4 The pictogram shows the number of DVDs owned by each of four friends.

(a) Who owns the most DVDs?
(b) How many more DVDs does Gerry own than Harry?

AQA

8 Fractions

This work will help you
- identify a fraction of a shape
- work out a fraction of a number

You need sheets FT–5 and FT–6.

A Recognising fractions level 4

- Which rectangles have one half shaded?
- What fraction is shaded in the others?
- Are there any other ways to shade $\frac{1}{2}$ of this rectangle?
- Can you shade $\frac{1}{4}$ of the rectangle?
- How about $\frac{3}{4}$ of it?

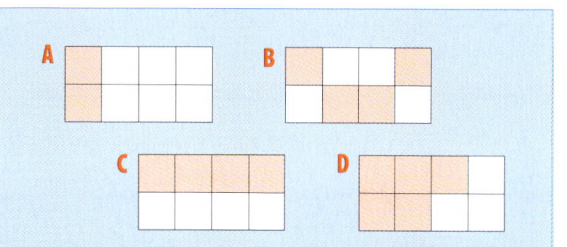

A1 Is a quarter of each of these squares shaded?
If not, say what fraction of the square is shaded.

(a) 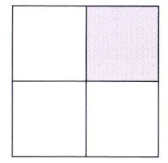 (b) 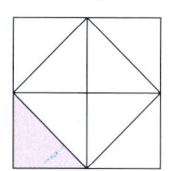 (c) 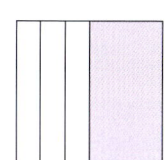 (d)

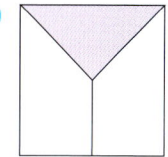

A2 What fraction of each of these shapes is shaded?

(a) 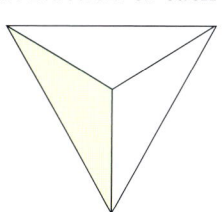 (b) 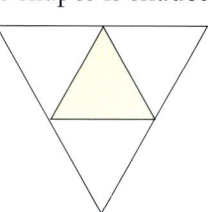 (c) 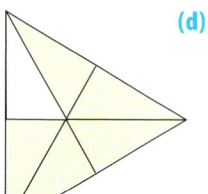 (d)

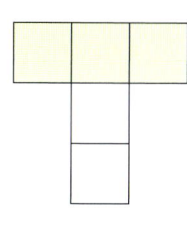

(e) 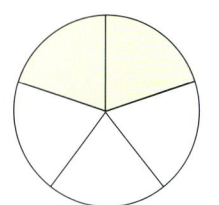 (f) 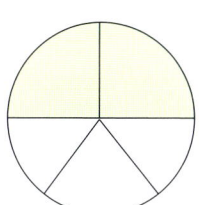 (g) 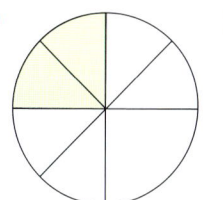 (h)

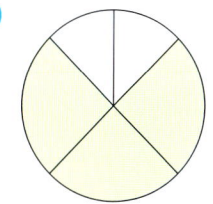

8 Fractions 39

A3 Make a copy of this rectangle on squared paper and shade $\frac{1}{4}$ of it.

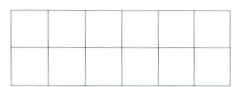

A4 Repeat A3 for the following fractions.
(a) $\frac{3}{4}$ (b) $\frac{1}{3}$ (c) $\frac{2}{3}$ (d) $\frac{1}{6}$ (e) $\frac{5}{6}$

B Finding a fraction of a number

B1 (a) Copy and complete this sentence.
   To find $\frac{1}{2}$ of a number divide it by …
   (b) Work out
   (i) $\frac{1}{2}$ of 30   (ii) $\frac{1}{2}$ of 36   (iii) $\frac{1}{2}$ of 68   (iv) $\frac{1}{2}$ of 280   (v) $\frac{1}{2}$ of 150

B2 (a) Copy and complete this sentence.
   To find $\frac{1}{4}$ of a number divide it by …
   (b) Work out
   (i) $\frac{1}{4}$ of 24   (ii) $\frac{1}{4}$ of 44   (iii) $\frac{1}{4}$ of 100   (iv) $\frac{1}{4}$ of 120   (v) $\frac{1}{4}$ of 180

B3 Work these out.
(a) $\frac{1}{3}$ of 15 (b) $\frac{1}{5}$ of 20 (c) $\frac{1}{8}$ of 16 (d) $\frac{1}{10}$ of 90 (e) $\frac{1}{7}$ of 21

B4 How many minutes are there in
(a) half an hour (b) quarter of an hour (c) one third of an hour

---

$\frac{3}{4}$ of 20 is the same as 3 lots of $\frac{1}{4}$ of 20.

$\frac{1}{4}$ of 20 = 5

So $\frac{3}{4}$ of 20 = 3 × 5 = 15

• Explain how you could find $\frac{4}{5}$ of 20.

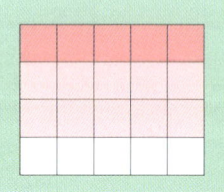

B5 (a) (i) Work out $\frac{1}{3}$ of 18.   (ii) Work out $\frac{2}{3}$ of 18.
  (b) (i) Work out $\frac{1}{6}$ of 60.   (ii) Work out $\frac{5}{6}$ of 60.

B6 Work these out.
(a) $\frac{3}{4}$ of 36 (b) $\frac{2}{3}$ of 24 (c) $\frac{2}{5}$ of 45 (d) $\frac{5}{6}$ of 30 (e) $\frac{2}{7}$ of 28
(f) $\frac{2}{3}$ of 180 (g) $\frac{3}{5}$ of 400 (h) $\frac{3}{8}$ of 320 (i) $\frac{3}{10}$ of 150 (j) $\frac{3}{4}$ of 120

B7 The first $\frac{1}{2}$ of TANK is TA. The last $\frac{1}{3}$ of IRRITABLE is BLE.
Put them together and you get TABLE.

What words do these make?

(a) The first $\frac{1}{3}$ of TRIANGLES and the second $\frac{1}{2}$ of STICKS

(b) The first $\frac{2}{5}$ of PRECIPICES and the last $\frac{2}{5}$ of VICTORIOUS

(c) The first $\frac{1}{3}$ of DISCUSSED, the first $\frac{1}{4}$ of APPLICATIONS and the last $\frac{3}{4}$ of FEAR

(d) The first $\frac{1}{4}$ of THURSDAY, the last $\frac{1}{2}$ of STAIRS and the first $\frac{1}{3}$ of TYRANT

(e) The first $\frac{3}{8}$ of COMPUTER, the last $\frac{3}{8}$ of SAUCEPAN and the last $\frac{1}{4}$ of CONSERVATION

B8 Find the missing numbers.

(a) $\frac{1}{3}$ of ■ = 8   (b) $\frac{1}{■}$ of 12 = 3   (c) $\frac{1}{■}$ of 40 = 5   (d) $\frac{3}{4}$ of ■ = 18

Further activities

Fraction maze sheet FT–5

Start in the shaded circle.
Move to the neighbouring circle with the number closest to the one you are on. But don't visit the same circle twice!

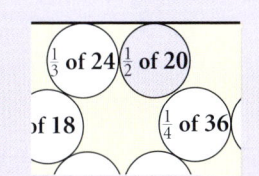

Continue until you reach the edge of the maze again.
Record where you go and the number you finish on.

$\frac{1}{3}$ of 24 | $\frac{1}{2}$ of 20 | of 18 | $\frac{1}{4}$ of 36

Fraction dominoes sheet FT–6

Use as many dominoes as you can to make a 'chain'.

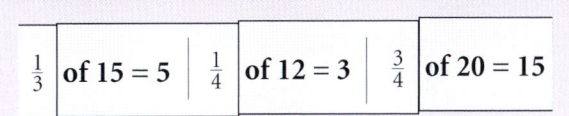

| $\frac{1}{3}$ of 15 = 5 | $\frac{1}{4}$ of 12 = 3 | $\frac{3}{4}$ of 20 = 15 |

Test yourself

T1 (a) Write down the fraction of this shape which is shaded.

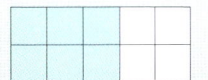

(b) Copy this shape.
Shade $\frac{3}{4}$ of the shape.

Edexcel

T2 Work these out.

(a) $\frac{1}{2}$ of 18   (b) $\frac{1}{4}$ of 20   (c) $\frac{3}{4}$ of 20   (d) $\frac{1}{5}$ of 150   (e) $\frac{3}{5}$ of 150

9 Decimal places

This work will help you

- find the position on a number line of a decimal with one or two decimal places
- put decimals with one or two decimal places in order of size

You need sheets FT–7, FT–8, FT–9 and FT–10.

A One decimal place   level 4

Arrow A is pointing to 3.2.
- What numbers are the other arrows pointing to?

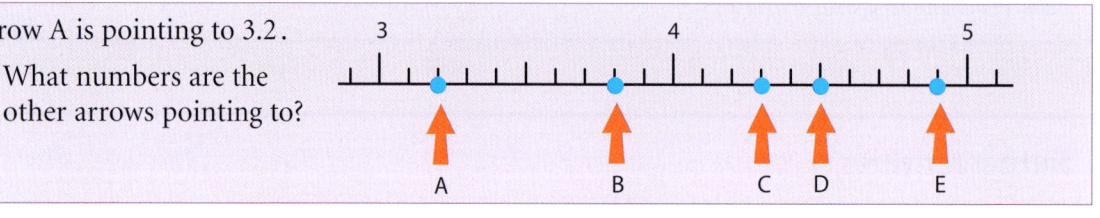

- Decide which of these are true and which are false.

  A 4 is smaller than 4.3
  B 4 is smaller than 3.9
  C 4 is smaller than 4.0
  D 4 is larger than 4.7
  E 4 is larger than 3.2
  F 4 is larger than 2.9

A1 What number does each arrow point to?

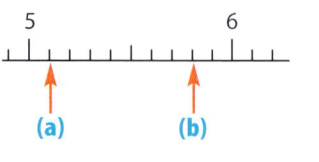

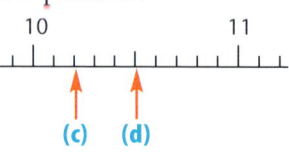

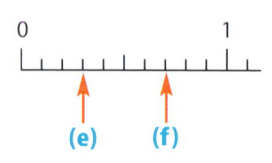

A2 Solve the puzzles on sheet FT–7.
Mark numbers and letters on the number lines to make words.

A3 What number does each arrow point to?

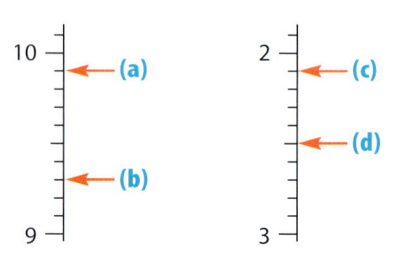

A4 Decide whether each of these is true or false.
(a) 5.6 is larger than 4.9
(b) 5 is larger than 4.8
(c) 3 is smaller than 3.9
(d) 3 is smaller than 2.1

A5 Which of the discs will go through the slot?

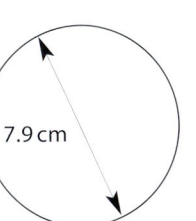

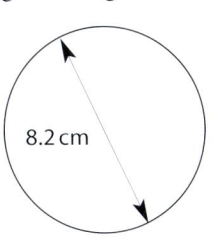

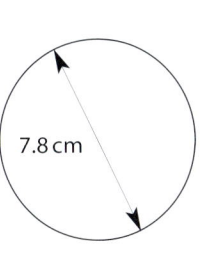

A 7.9 cm   B 8.2 cm   C 7.8 cm   8 cm

A6 Which of these numbers are between 2 and 4?

3.3, 2.3, 4.5, 3, 3.8, 6.3, 1.9

A7 Put each list of numbers in order, smallest first.

(a) 1.5, 5.4, 3.2, 4.9, 10.1
(b) 3.4, 7, 6.7, 1.2, 2
(c) 1.7, 1.2, 2, 0.9, 0.1
(d) 0.1, 1, 0.5, 0.2, 1.3

A8 In a guessing contest, some students guess the length of this line.
Their guesses are

6.5 cm   6.6 cm   7.1 cm   6 cm   7.5 cm

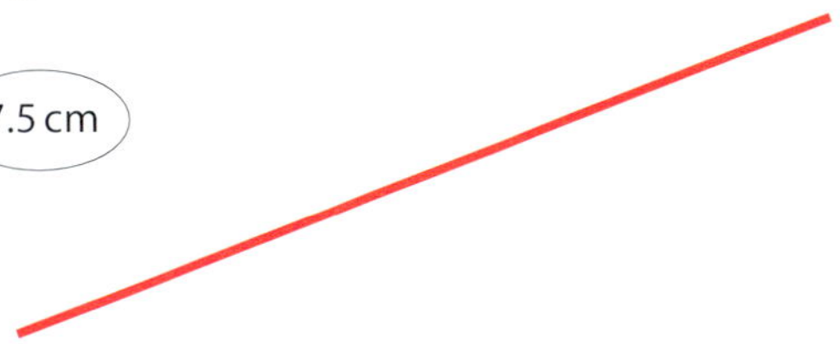

(a) Measure the length of the line.
(b) Which guess is the closest?

A9 What is the number halfway between 2 and 3?

A10 The graph shows the average monthly rainfall at Kew in London.
(J, F, M, … stand for January, February, March, …)

On average …

(a) What is the monthly rainfall at Kew for
  (i) January   (ii) September
  (iii) May   (iv) December

(b) Which are the three driest months?

(c) How much rain falls in the wettest month?

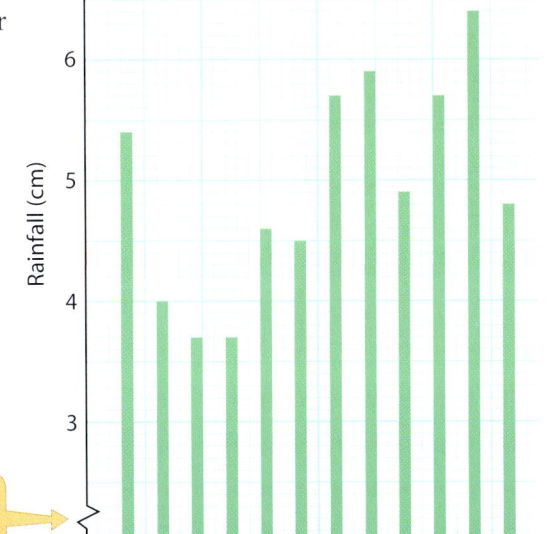

This tells you that the rainfall axis does not start at zero.

Average monthly rainfall at Kew
Rainfall (cm)
Month

B Two decimal places   level 4

The number lines below show numbers between 1 and 2.
The top line is divided into tenths.
The bottom line is divided into hundredths.

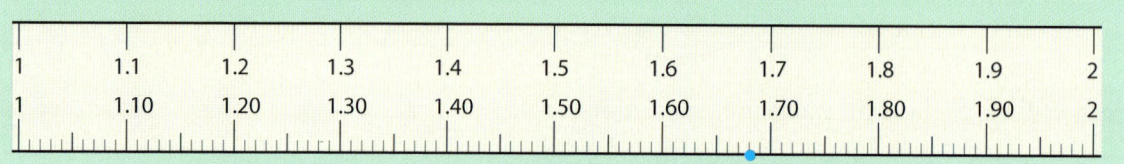

| 1 | 1.1 | 1.2 | 1.3 | 1.4 | 1.5 | 1.6 | 1.7 | 1.8 | 1.9 | 2 |
| 1 | 1.10 | 1.20 | 1.30 | 1.40 | 1.50 | 1.60 | 1.70 | 1.80 | 1.90 | 2 |

The number 1.68 is marked.
The diagram shows that 1.68 is between 1.6 and 1.7.
- Give two numbers that are between 1.3 and 1.4.
- Give two numbers that are between 1 and 1.1.

- Decide which of these are true and which are false.

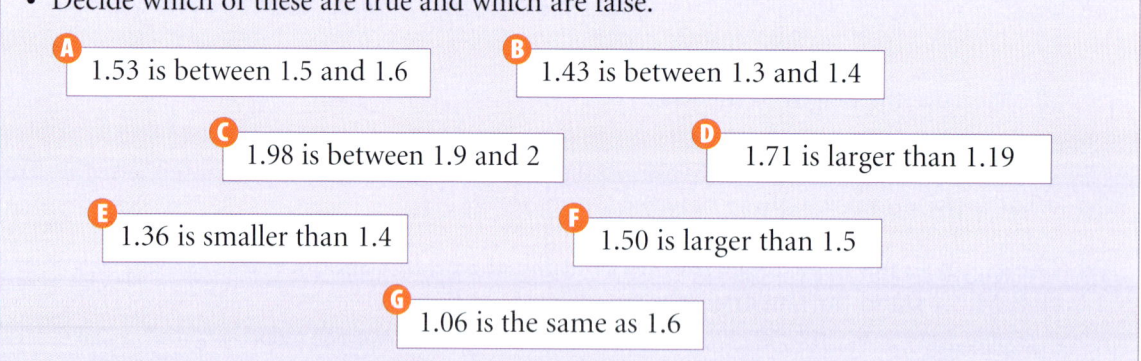

  A 1.53 is between 1.5 and 1.6

  B 1.43 is between 1.3 and 1.4

  C 1.98 is between 1.9 and 2

  D 1.71 is larger than 1.19

  E 1.36 is smaller than 1.4

  F 1.50 is larger than 1.5

  G 1.06 is the same as 1.6

B1 What number does each arrow point to?

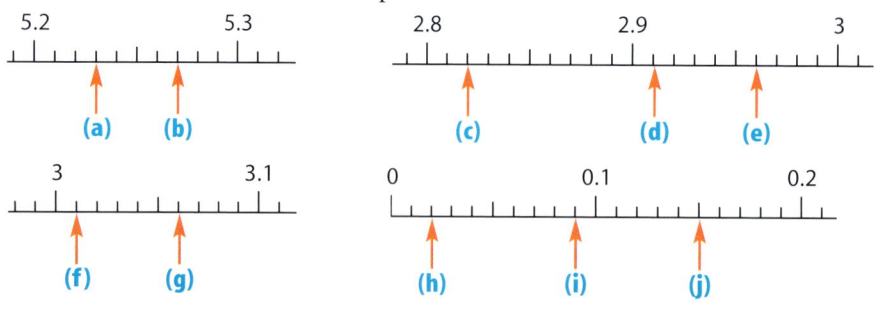

(arrows (a), (b) on line from 5.2 to 5.3)
(arrows (c), (d), (e) on line from 2.8 to 3)
(arrows (f), (g) on line from 3 to 3.1)
(arrows (h), (i), (j) on line from 0 to 0.2)

B2 Decide whether each of these is true or false.
(a) 6.49 is larger than 6.23
(b) 5.24 is larger than 5.3
(c) 2 is larger than 1.35
(d) 3.09 is smaller than 3.1

B3 Solve the puzzles on sheet FT–8.
Make words by putting numbers in order.

B4 Here are some cars.

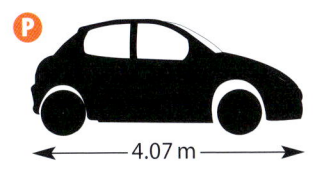

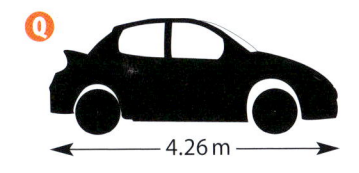

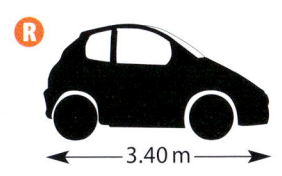

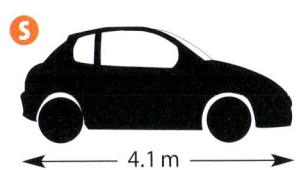

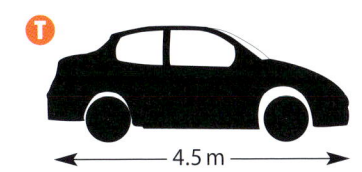

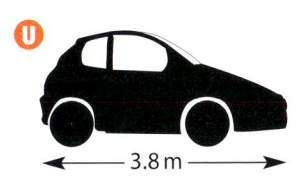

P — 4.07 m
Q — 4.26 m
R — 3.40 m
S — 4.1 m
T — 4.5 m
U — 3.8 m

The garages below are both wide enough for the cars but they may not be long enough.

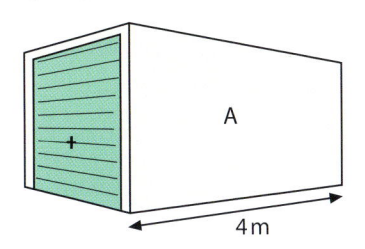

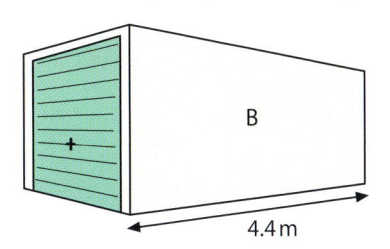

Garage A — 4 m
Garage B — 4.4 m

(a) Which cars could go in garage A? (b) Which could go in garage B?

B5 Solve the puzzles on sheet FT–9.
Make words by putting numbers in order.

B6 Which of these numbers are between 2.5 and 2.7?

3.6, 2.6, 2.67, 2.52, 2.06

B7 Put these numbers in order, smallest first.

7.1, 7.92, 8, 7.4, 7.28

B8 Put these numbers in order, **largest** first.

0.3, 1.2, 1.03, 0.06

B9 In a guessing contest, some students guess the weight of a cat.
Their guesses are

3.7 kg   3.85 kg   3.95 kg   3.8 kg   3.9 kg

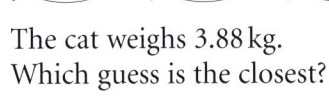

The cat weighs 3.88 kg.
Which guess is the closest?

B10 What number is halfway between

(a) 3 and 4   (b) 3.2 and 3.3   (c) 9.7 and 9.8   (d) 1 and 1.1

***B11** Find your way through the 'closest neighbour maze' on sheet FT–10.
The instructions for the maze are on the sheet.

9 Decimal places 45

C Decimal lengths   level 4

This is the plan of the ground floor of Gill's house.

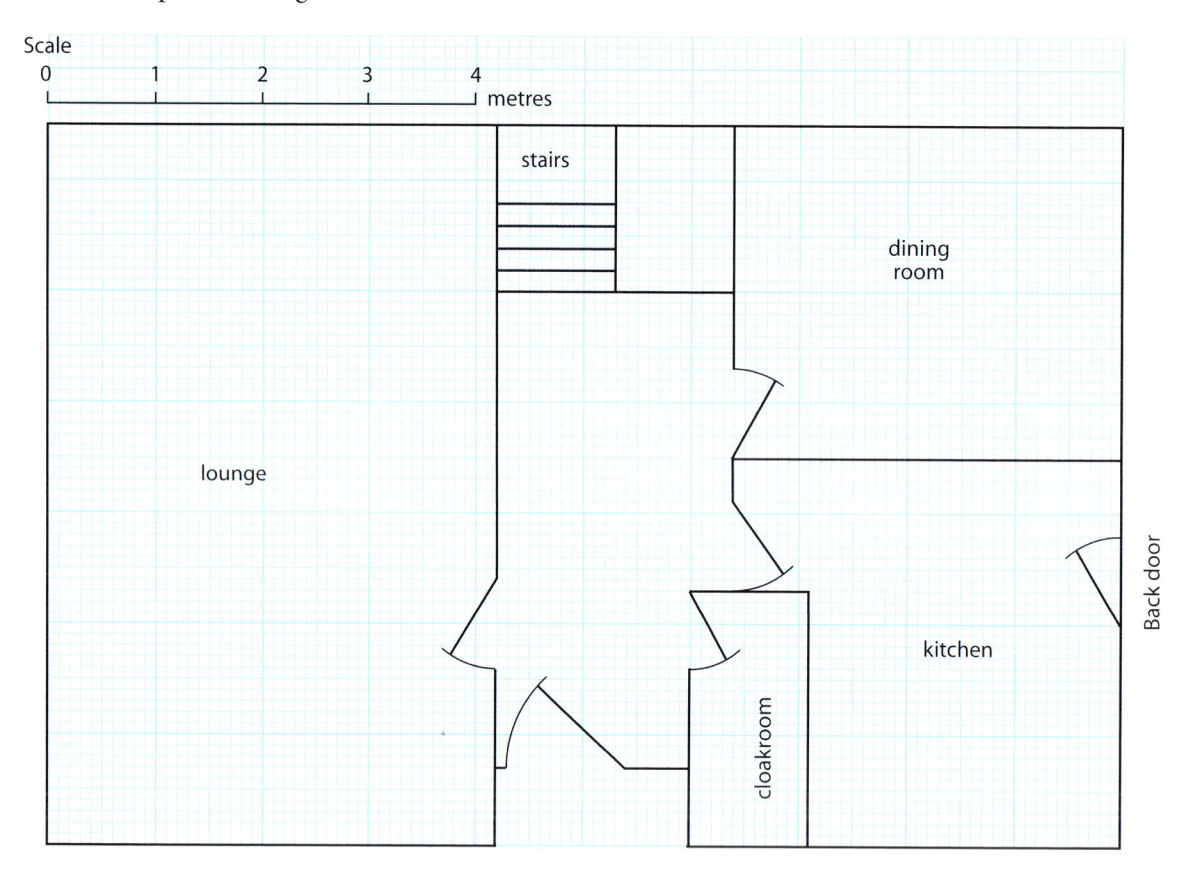

C1 (a) What is the width of the lounge?
  (b) What is the length of the lounge?

C2 (a) What is the width of the cloakroom?
  (b) What is the length of the cloakroom?
  (c) Gill buys a coat rack that is 1.85 metres long.
      Will it fit in the cloakroom?

C3 (a) What is the width of the dining room?
  (b) What is the length of the dining room?
  (c) Gill buys a large rug on holiday.
      It measures 2.9 metres by 3.35 metres.
      Will it fit in the dining room?

C4 (a) How wide are the stairs?

(b) Which of the stair carpets below are too wide for the stairs?

1.5 m   1.05 m   1.15 m

1.25 m   1 m   0.95 m

C5 A kitchen cupboard is 1.5 m long, 0.75 m wide and 1.2 m high.
Will it fit through the back door?

C6 A sofa is 0.84 metres wide.
Will it fit through the door of the lounge without any twisting or turning?

Test yourself

T1 What number does each arrow point to?

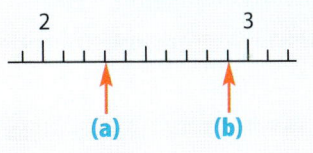

T2 Write down three different numbers that are between 4 and 5.

T3 Which arrow is pointing to 5.02?

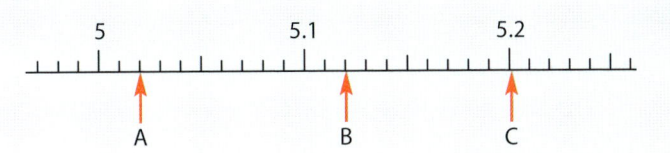

T4 Mary has six cats.
She weighs them in kilograms.
She writes their weights in order as shown below.

One of the weights is in the wrong place.

    3.3,   3.12,   3.24,   3.31,   3.52,   3.62

Which weight is in the wrong place?

T5 Put these numbers in order, smallest first

    2.36   2.1   2.09   2.7

10 Median and range

This work will help you

- find the median and range of a set of data
- use the median and range to compare two data sets

A Median (for an odd number of data items)

Jane keeps hens that lay eggs.
Last week one of the hens laid nine eggs.
Their weights in grams were

    67 72 55 81 77 60 58 77 66

Jane writes these weights in order, smallest first:

    55 58 60 66 (67) 72 77 77 81

Jane has put a ring round the middle weight in the list.
This weight is called the **median** weight of the eggs.

The median weight, 67 g, is an 'average' weight.
Some eggs weigh more than 67 g and some less, but 67 g is 'in the middle'.

Writing a list of numbers in order

Suppose you want to write this list in order: 48 36 42 39 47 39 50

First write the list as given.
Find the smallest, cross it off and write it down on the line below.

    48 ~~36~~ 42 39 47 39 50
    36

Now find the next smallest.
It occurs twice, so write it down twice on the line below.

    48 ~~36~~ 42 ~~39~~ 47 ~~39~~ 50
    36 39 39

And so on.

A1 Here are the weights in grams of a set of seven eggs.

    62 54 65 53 75 64 70

(a) Write the weights in order, smallest first.

(b) Write down the median weight.

A2 Write each of these lists in order, smallest first.
Then write down the median.

(a) 49 44 51 46 50   (b) 28 63 44 50 29 47 51 37 33

(c) 43 65 43 38 51 63 55   (d) 7 6 7 2 5 9 3 5 6 4 3

B Median (for any number of data items)

One of Jane's hens laid ten eggs last week. Here are the weights in grams, already in order.

56 57 61 61 62 66 69 71 76 82

This time there is no single 'middle weight', because 10 is an even number.
So Jane puts a ring round the middle pair of weights.

56 57 61 61 ⬚62 66⬚ 69 71 76 82

The median weight is halfway between the middle pair.
You can find it by adding the middle pair together and then dividing by 2.

62 + 66 = 128   128 ÷ 2 = 64

The median weight is 64 g.

B1 These eight weights in grams are in order.

49 53 57 58 62 65 69 73

Write down the middle pair and then find the median weight.

B2 (a) Write these six weights in order.

54 71 80 67 59 63

(b) Pick out the middle pair and find the median weight.

B3 These ten weights are in order.

47 50 52 53 53 56 59 63 67 72

Pick out the middle pair and find the median weight. (It is not a whole number.)

B4 (a) Write these six weights in order.   62 67 71 64 58 67

(b) Find the median weight.

B5 Find the median of each of these sets of weights in grams.

(a) 18 12 13 19 16 13 14 21
(b) 38 51 63 80 66 54 47
(c) 96 105 92 90 103 99
(d) 25 31 29 30 22 27 33 28

B6 (a) Last week a hen laid seven eggs with these weights in grams.

60 78 61 56 74 59 58

Find the median weight of these eggs.

(b) This week the same hen laid ten eggs with these weights in grams.

72 62 60 58 77 54 57 63 76 71

Find the median weight of these eggs.

(c) Did the median weight go up or go down between last week and this week?

C Range

The **range** of a set of data is the difference between the largest and smallest values.

Here is a set of weights in grams.   63   71   59   68   52   55

The largest value is 71 and the smallest is 52.

So the range of the weights is 71 − 52 = 19 g.

The range tells you how spread out the weights are.

C1 For each of these sets of weights in grams, find

   (i) the largest value   (ii) the smallest value   (iii) the range

   (a) 44   48   53   39   64   71   53   89   67
   (b) 620   540   480   490   630   610   560   600
   (c) 57   49   63   82   54   48   55   60   74   79   81   80   64
   (d) 552   671   504   622   673   650

C2 Rajesh measures the heights of the male students in his class.
   The heights in centimetres are

   174   168   177   159   171   166   162

   (a) Write the heights in order, smallest first.
   (b) Find the median height.
   (c) Find the range of the heights.

C3 Kamila measures the heights of the female students in her class.
   The heights in centimetres are

   161   159   167   163   166   164   166   168

   (a) Find the median height.
   (b) Find the range of the heights.

C4 Find the median and range of each of these data sets.

   (a) 62   81   53   47   54   60   73
   (b) 52   73   82   65   57   65   52   65
   (c) 33   19   34   41   23   59   47   48   52

C5 The ages of the members of a sports club, starting with the youngest, are

   15   17   18   18   19   20   20   22   25   27   30   30   31   33   39   44

   (a) Find the median age.
   (b) Find the range of the ages.

   A new member joins the club. Her age is 46.

   (c) Find the median age of the club after the new member has joined.
   (d) Find the range after the new member has joined.

D Comparing two sets of data

The word **data** is used for any set of recorded numbers, weights, heights and so on.

The median and the range are often used to compare two sets of data.

Here are the weights in grams of the eggs laid by two hens A and B.
Each set of data has already been written in order.

   Hen A   51   54   57   58   60   67   68   69   71
   Hen B   57   58   58   59   61   65   67   69   70   73

Comparing medians

The median weight for hen A is 60 g.
The median weight for hen B is halfway between 61 and 65, which is 63 g.

So, on average, hen B's eggs are heavier than hen A's.

Comparing ranges

The range for hen A is $71 - 51 = 20$ g.
The range for hen B is $73 - 57 = 16$ g.

So hen A's eggs are more widely spread out in weight than hen B's.

D1 The ages of the members of two football teams A and B are as follows.

   Team A   27   23   19   26   23   18   29   31   26   24   20
   Team B   24   21   28   20   34   30   27   21   31   19   30

(a) Find the median age of team A.

(b) Find the median age of team B.

(c) Which team is older, on average?

(d) Find the range of the ages of team A.

(e) Find the range of the ages of team B.

(f) In which team are the ages more spread out?

D2 Brian's class consists of 11 girls and 12 boys.
Brian records the weight, in kilograms, of each boy and each girl.
Here are the two data sets.

   Girls   42   35   39   38   44   36   33   40   35   44   37
   Boys   47   34   48   40   43   38   36   44   48   37   35   35

(a) Find   (i) the median of the girls' weights   (ii) the range of the girls' weights

(b) Find   (i) the median of the boys' weights   (ii) the range of the boys' weights

(c) Which of the two groups, girls or boys, is heavier on average?

(d) Which of the two groups is more spread out in weight?

D3 Marsha is doing an experiment on handwriting.
She asks some right-handed students to write the alphabet using their right hand.
She times each student. Here are the times in seconds.

    18 16 17 24 28 22 25 20 17 21

(a) Find   (i) the median time   (ii) the range of the times

Marsha now asks the students to write the alphabet with their left hand.
Here are the times in seconds.

    38 35 42 39 50 47 41 36 49 48

(b) Find   (i) the median time   (ii) the range of the times

(c) Which set of times was more spread out, the right-hand times or the left-hand times?

D4 Jack gave a group of boys a message to text and timed them sending it.
The times taken, in seconds, were

    24 21 28 30 25 23 27 33 27 28

(a) Find   (i) the median time   (ii) the range of the times

A group of girls was given the same message to text.
Their median time was 25 seconds and the range of their times was 15 seconds.

(b) Which group, boys or girls, took longer on average?

(c) For which group were the times more spread out?

Test yourself

T1 Find the median of each of these sets of data.

(a) 44 48 53 39 64 71 53 89 67

(b) 27 19 52 24 18 25 30 44 49

T2 Find the range of this set of data.   27 21 28 39 42 26 22 38 36

T3 The temperatures in °C at eight places in Yorkshire on a certain day were

    15 13 12 14 17 13 15 16

(a) Find the median temperature in Yorkshire.

(b) Find the range of the temperatures in Yorkshire.

On the same day, the temperatures in nine places in Wales were

    14 13 15 12 16 10 12 15 13

(c) Find the median temperature in Wales.

(d) Find the range of the temperatures in Wales.

(e) Which area, Yorkshire or Wales, was warmer on average?

(f) In which area were the temperatures more spread out?

11 Mental methods 2

You will practise multiplying and dividing by 10, 100, 1000.

This work will help you multiply by numbers like 20, 300, …

A Multiplying by 10, 100, 1000

Multiplying by 10, 100, 1000, … involves moving digits to the left.

387 × 10 = 3870   2.84 × 100 = 284   1.8 × 1000 = 1800

Move one place left. The 0 shows there are no units.

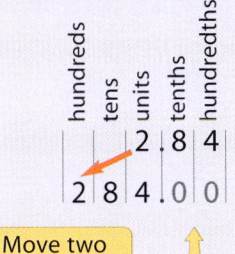

Move two places left. You could put zeros in, but you don't need them here.

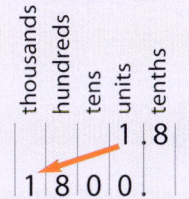

Move three places left. There are no tens and no units.

A1 Work these out in your head.
  (a) 5 × 10   (b) 5 × 100   (c) 5 × 1000   (d) 2.5 × 10   (e) 2.5 × 100

A2 Write down the answers to these.
  (a) 25 × 10   (b) 432 × 100   (c) 10 × 650   (d) 72 × 1000   (e) 8000 × 10
  (f) 2.45 × 10   (g) 3.2 × 100   (h) 42.1 × 100   (i) 0.2 × 1000   (j) 0.58 × 100

A3 A 5p coin weighs 3.25 g.
What is the weight of one hundred 5p coins?

A4 (a) A 20p coin is 1.7 mm thick.
How tall is a pile of ten 20p coins?

  (b) A 50p coin is 1.78 mm thick.
How tall is a pile of ten 50p coins?

B Dividing by 10, 100, 1000

Dividing by 10, 100, 1000, … involves moving digits to the right.

420 ÷ 10 = 42

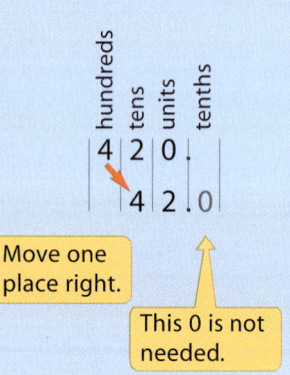

hundreds	tens	units	tenths
4	2	0	.
	4	2	. 0

Move one place right.

This 0 is not needed.

521 ÷ 100 = 5.21

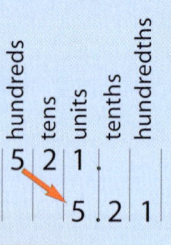

hundreds	tens	units	tenths	hundredths
5	2	1	.	
		5	. 2	1

Move two places right.

- How would you work out 673 ÷ 1000?

B1 Work these out in your head.
(a) 400 ÷ 10 (b) 400 ÷ 100 (c) 400 ÷ 1000 (d) 62 ÷ 10 (e) 62 ÷ 100

B2 Write down the answers to these.
(a) 18 ÷ 10 (b) 5610 ÷ 100 (c) 3400 ÷ 1000 (d) 764 ÷ 100 (e) 43 ÷ 10
(f) 25 ÷ 100 (g) 360 ÷ 1000 (h) 4.5 ÷ 10 (i) 97 ÷ 100 (j) 0.3 ÷ 10

B3 Ben buys a pack of 10 pens for £4.50. What is the cost of one pen?

B4 A bag of 100 wood screws weighs 275 g. What is the weight of one wood screw?

B5 (a) A pack of 100 blank CDs costs £14. What is the cost of one CD?

(b) The thickness of one CD is 1.3 mm. The CDs are packed in a single pile. What is the height of the pile of 100 CDs?

C Multiplying by numbers ending in zeros

30 is 3 × 10.
To multiply by 30, you can multiply by 3 and then by 10.

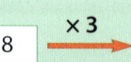

8 × 30   8 →×3→ 24 →×10→ 240

- How would you work out 8 × 300, 8 × 3000, …?

54   11 Mental methods 2

C1 (a) 2×80 (b) 4×200 (c) 7×300 (d) 6×40 (e) 9×50
(f) 5×30 (g) 6×90 (h) 8×200 (i) 3×900 (j) 2×70

To work out 30×400 …   Start with 3×4.
From this you can work out 30×4.
Then you can do 30×40, …
… and 30×400.

$3 \times 4 = 12$
$30 \times 4 = 120$
$30 \times 40 = 1200$
$30 \times 400 = 12\,000$

C2 (a) 2×60 (b) 20×60 (c) 20×600 (d) 200×60 (e) 200×600

C3 (a) 30×600 (b) 700×200 (c) 20×900 (d) 60×60 (e) 300×50

C4 (a) 500×40 (b) 80×60 (c) 50×700 (d) 2000×40 (e) 60×5000

C5 Felt-tips cost 30p each.
Work out the cost, in pounds, of 200 felt-tips.

C6 Work out the cost, in pounds, of
(a) 80 rulers at 40p each
(b) 500 pencils at 20p each
(c) 400 pens at 60p each

C7 A football stadium has 4000 seats.
Tickets are sold at £20 each.
How much money is made if all the tickets are sold?

Test yourself

T1 Write down the answers to these.
(a) 92×10 (b) 1000×31 (c) 40.5×10 (d) 100×2.6 (e) 0.68×100
(f) $420 \div 10$ (g) $8000 \div 100$ (h) $79 \div 100$ (i) $4.3 \div 10$ (j) $0.5 \div 10$

T2 Work these out.
(a) 60×30 (b) 200×40 (c) 5000×80 (d) 400×90 (e) 600×700

T3 (a) Mary bought 8 strawberry plants. She paid 70p for each plant.
Work out the total cost of 8 plants.
Give your answer in pounds.
(b) Peter bought some raspberry canes. He paid £16 for 10 canes.
Work out the cost of one cane.

OCR

12 Solids, nets and views

This work will help you
- recognise simple solids and their nets
- count the numbers of faces, edges and vertices a solid has
- decide what parts of a solid you can see from different viewpoints

You may need sheet FT–11.

A Solids and nets   level 4

- Match up each solid with its mathematical name.

  Cube   Cuboid   Cylinder   Cone   Sphere   Prism   Pyramid

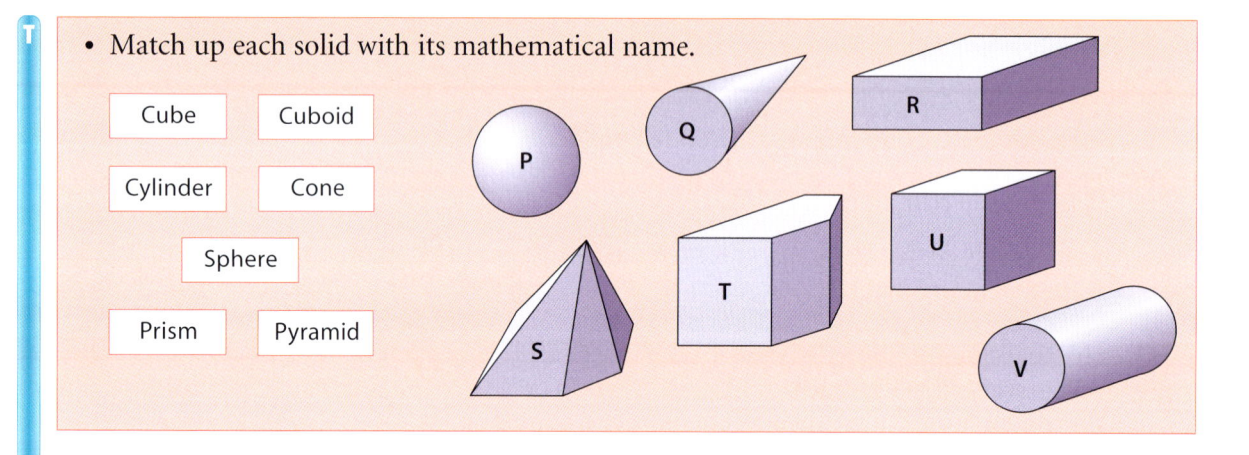

  P, Q, R, S, T, U, V

In diagrams of solids, hidden edges are sometimes shown with dotted lines.

This pyramid has
- 5 faces
- 5 vertices
- 8 edges

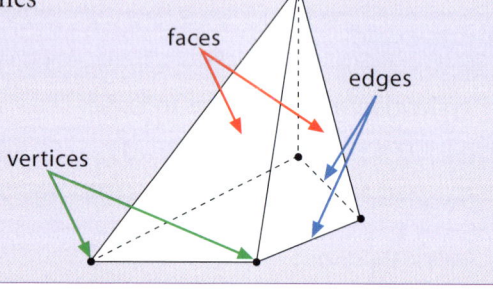

A1 For each solid below (i) give its mathematical name
  (ii) state the numbers of faces, vertices and edges.

(a)   (b) 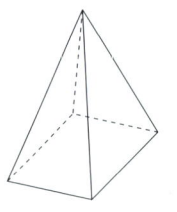   (c)

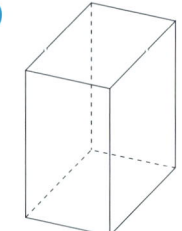

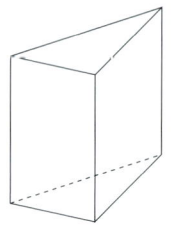

A **net** is a flat 2-dimensional shape that can be folded to make a solid 3-dimensional shape.

You can check your answers to A2 and A3 by cutting out and folding the nets on sheet FT–11.

A2 Which of these are nets that fold up to make a cube?

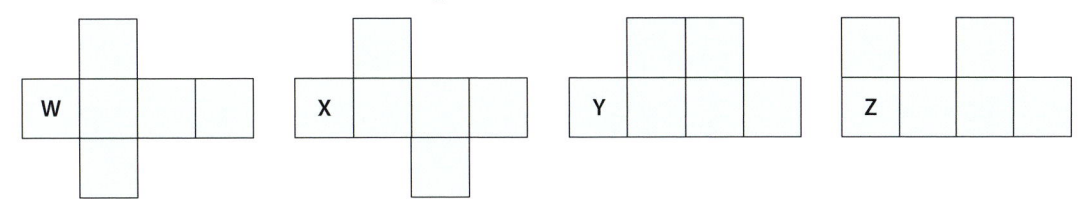

W   X   Y   Z

A3 State whether each net below gives a **cuboid**, a **pyramid** or a **prism**.

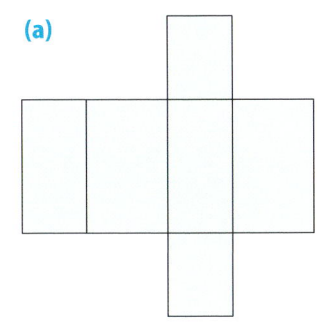

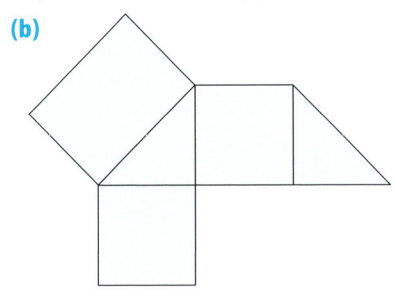

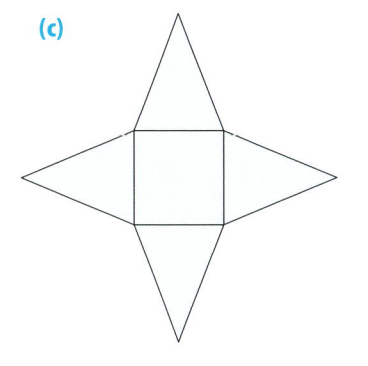

(a)   (b)   (c)

A4 Copy each diagram below and add one square to make each one the net of a cube.

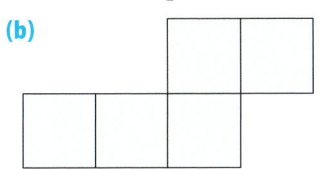

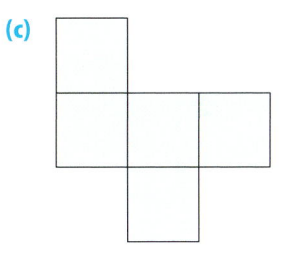

(a)   (b)   (c)

A5 (a) What is the name of the solid that can be made from this net?
Choose a name from this list.

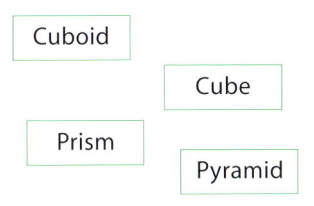

Cuboid
Cube
Prism
Pyramid

(b) How many faces does the solid have?
(c) How many vertices does the solid have?

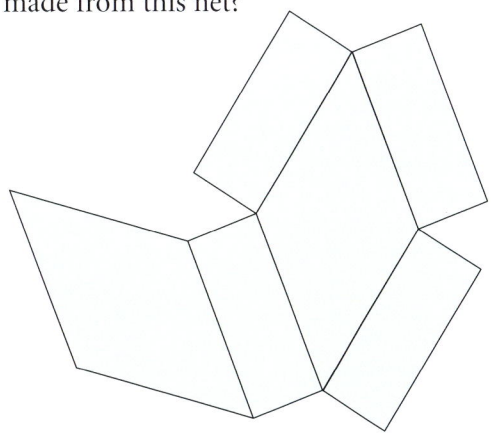

12 Solids, nets and views 57

B Views

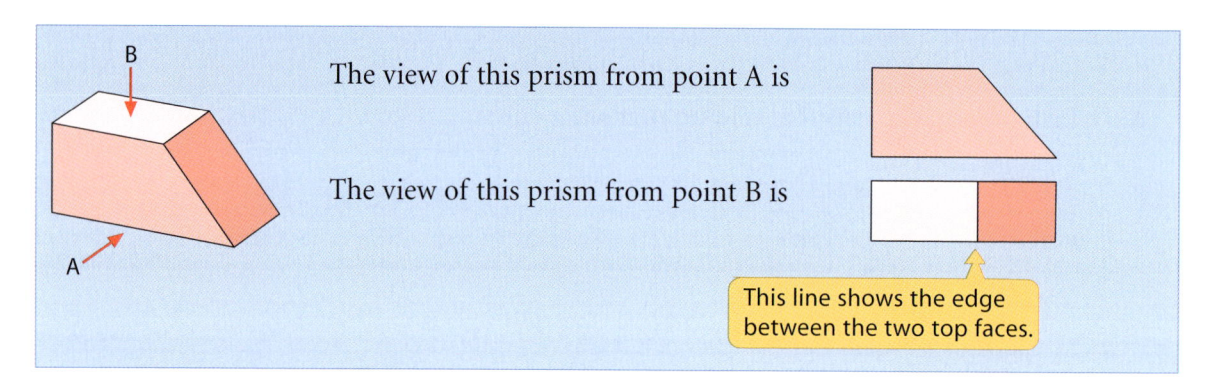

The view of this prism from point A is

The view of this prism from point B is

This line shows the edge between the two top faces.

B1 Below are five views of this prism, looking from the directions P, Q, R, S and T. Match each view with its correct letter.

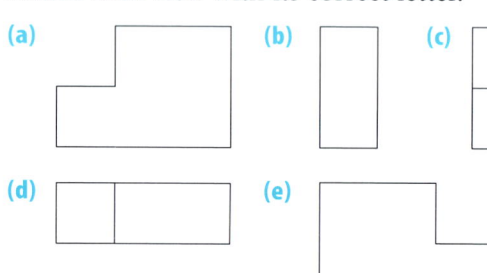

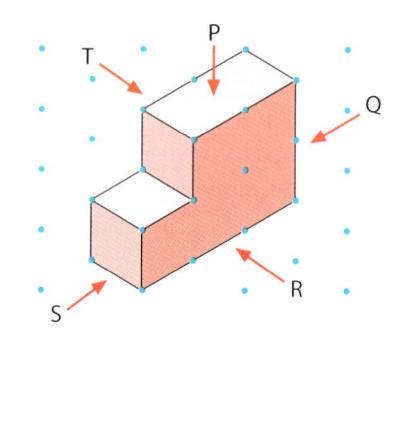

(a) (b) (c)

(d) (e)

B2 The diagrams show the net of a cube and this same cube in different positions.

What is on each face, looking in the direction of the arrow?

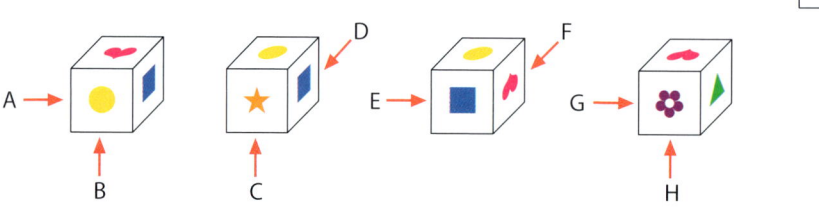

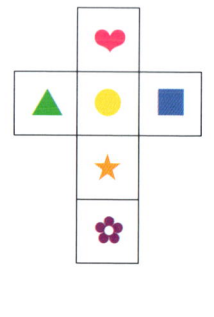

B3 Which of the diagrams is the view of the square-based pyramid from the top?

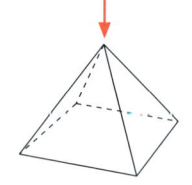

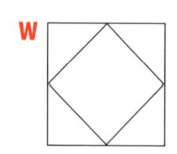

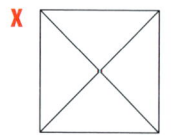

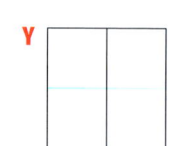

W   X   Y   Z

Test yourself

T1 Write down the mathematical name for each of these three different 3-D shapes.

(a)   (b)   (c)

Edexcel

T2 Look at these shapes.
Write **Yes** for the shapes that are the nets of a cuboid.
Write **No** for those that **are not** the nets of a cuboid.

(a)   (b)   (c)   (d)

OCR

T3

(a) What is the name of the solid that can be made from this net?
Choose a name from this list.

Cuboid   Pyramid   Cylinder   Prism   Sphere

(b) How many faces does the solid have?
(c) How many vertices does the solid have?

OCR

T4 Which of the diagrams is the view of the prism from the top?

P   Q   R

12 Solids, nets and views   59

13 Weighing

This work will help you use grams and kilograms.

A Grams and kilograms

The **weight** (or strictly speaking the **mass**) of a brass drawing pin is about one **gram**.
- Can you suggest other things that weigh about a gram?

A **kilogram** is a thousand grams.
- Can you suggest things that weigh about a kilogram?

A1 What completes each statement, grams or kilograms?
  (a) Joe's apple weighs 80 _____ .
  (b) Joe weighs 80 _____ .
  (c) My dictionary weighs 1600 _____ .

A2 Lou buys a bag of pasta that weighs 2 kilograms.
How many grams of pasta is this?

---

600 grams can be written as 600 g.
6 kilograms can be written as 6 kg.

A3 How many grams are there in each of these?
  (a) 3 kg
  (b) 9 kg
  (c) 10 kg
  (d) $\frac{1}{2}$ kg
  (e) $1\frac{1}{4}$ kg

A4 Sue is cooking chocolate brownies.
She uses 150 grams of flour from a 1 kg bag.
How much flour is left in the bag?

A5 How many of each of these packets would be needed to make 1 kg?

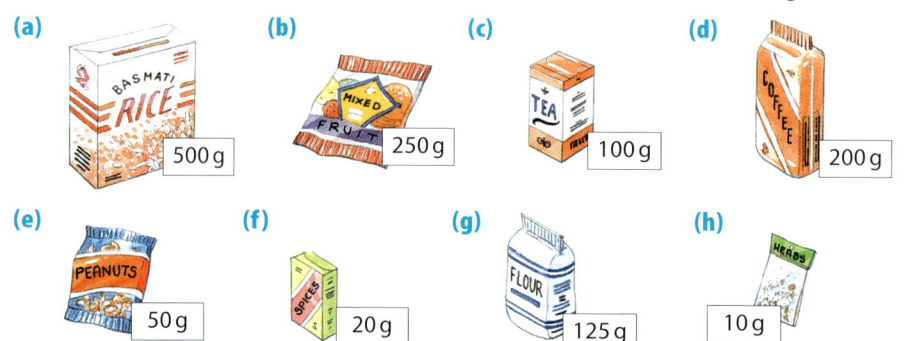

(a) RICE 500 g
(b) MIXED FRUIT 250 g
(c) TEA 100 g
(d) COFFEE 200 g
(e) PEANUTS 50 g
(f) SPICES 20 g
(g) FLOUR 125 g
(h) HERBS 10 g

A6 What is the weight in kilograms of 8 bags of rice that each weigh 500 g?

B Using decimals

- How many grams are in 1.5 kg?
- How would you convert 2650 grams to kilograms?

B1 How many grams are there in each of these?

(a) 0.5 kg (b) 3.5 kg (c) 0.4 kg (d) 0.75 kg
(e) 3.2 kg (f) 4.25 kg (g) 1.36 kg (h) 1.45 kg

B2 Wholesome Catering Company is making a large batch of banana cake. Here are the ingredients.

Copy the list but change all the amounts in grams to kilograms.

Banana cake (makes 100 slices)
4500 g ripe bananas (mashed)
500 g chopped nuts
1000 g soya margarine
1250 g raisins
750 g rolled oats
1500 g wholewheat flour

B3 Harry has a bag of rice that weighs 2.5 kg.
He cooks 800 g of it.
How much rice does he have left?

B4 Put these weights in order, smallest first.   1200 g   1.5 kg   200 g   0.7 kg   7 g

Test yourself

T1 How many grams are in 6 kg?

T2 Safiq is baking a cake.
He opens a $\frac{1}{2}$ kg bag of flour.
He uses 50 g of the flour.
How much flour is left?

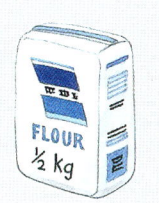

OCR

T3 Write the weights of these items in kilograms.

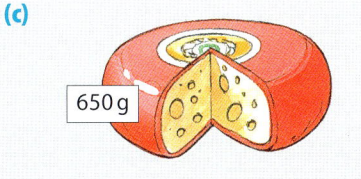

(a) 2000 g (Frozen Peas)   (b) 1250 g (Pasta)   (c) 650 g (cheese)

T4 A piece of cheese weighs 1.6 kg. Write the weight of the cheese in grams.

T5 What is the weight in kilograms of 6 packs of butter that each weigh 250 g?

Review 2

1 Sue notes the birds that visit her garden one morning. Here is her tally table.

Bird	Tally
Blackbird	IIII
Starling	IIII IIII I
Sparrow	III
Chaffinch	IIII III
Blue tit	IIII IIII III
Robin	IIII I

   (a) How many blackbirds did she see?
   (b) What was the frequency of robins?
   (c) What kind of bird is the mode?
   (d) Draw a frequency chart for the data.
   (e) How many birds did she see altogether?

2 Work out $\frac{3}{4}$ of 48.

3 The ages of the members of a volleyball team are
   32   28   25   42   31   35
   (a) Find the median age.
   (b) Find the range of these ages.

4 Write down the answers to these.
   (a) 3.2×100
   (b) 8.91×10
   (c) $500 \div 10$
   (d) $650 \div 100$
   (e) 500×70

5 (a) What weight is halfway between 1.2 kg and 1.3 kg?
   (b) How many grams are in $\frac{1}{4}$ kg?

6 Write down the mathematical name for this 3-D shape.

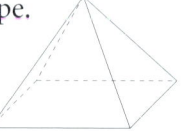

7 Write these weights in order, smallest first.
   1250 g $1\frac{1}{2}$ kg   1.9 kg   800 g

8 Which of the diagrams is the view of the prism from A?

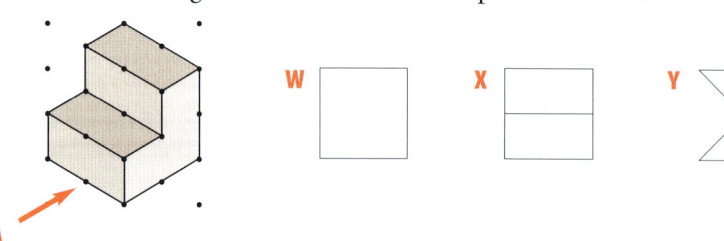

   W   X   Y   Z

9 Write this list of numbers in order, smallest first.
   3.6   4   1.3   1.08   3.58

14 Time and travel

This work will help you
- work out starting times, finishing times and time intervals
- read timetables

You need sheets FT–12 and FT–13, and a dice.

A Understanding 12-hour and 24-hour clock time   level 3

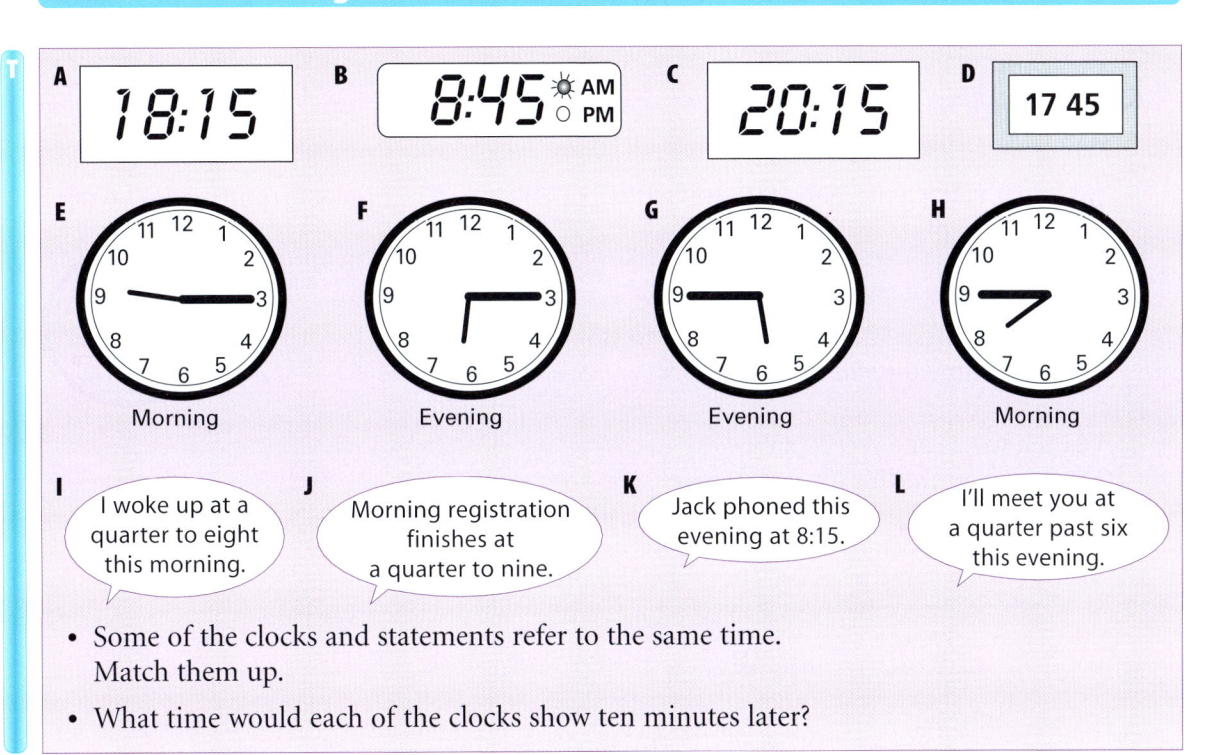

A 18:15   **B** 8:45 AM/PM   **C** 20:15   **D** 17 45

E (clock) Morning   **F** (clock) Evening   **G** (clock) Evening   **H** (clock) Morning

I I woke up at a quarter to eight this morning.
J Morning registration finishes at a quarter to nine.
K Jack phoned this evening at 8:15.
L I'll meet you at a quarter past six this evening.

- Some of the clocks and statements refer to the same time. Match them up.
- What time would each of the clocks show ten minutes later?

A1 Change these times into 24-hour clock times.
- (a) 3:40 p.m.
- (b) half past midnight
- (c) twenty to two in the morning
- (d) 9:15 p.m.
- (e) 8:40 a.m.
- (f) 10 minutes before midday

A2 Change these times into 12-hour clock times using a.m. and p.m.
- (a) 09:00
- (b) 12:30
- (c) 16:45
- (d) 20:10
- (e) 23:40

A3 Put these times in order, earliest first.

23:15   7:45 a.m.   2:00 p.m.   03:05   17:35   7:15 p.m.

B Time intervals
level 3

> Morning break starts at 10:40 and ends at 11:05. How long is break?

> The film starts at half past eight. It is 1 hour 40 minutes long. What time does it end?

B1 (a) Copy and complete this diagram.
(b) How long is it from 15:45 to 16:10?

B2 How long is it
(a) from 14:55 to 15:05
(b) from 11:30 to 12:15
(c) from 7:50 a.m. to 8:05 a.m.
(d) from 06:45 to 07:30
(e) from 19:15 to 20:10
(f) from 4:55 p.m. to 5:40 p.m.

B3 (a) The clock shows the time when Mike leaves the house in the morning. What time does it show?
(b) It takes Mike 45 minutes to drive to work. What time does he arrive at work?

B4 Clara leaves home at 9:20 a.m.
She arrives at the library at 9:50 a.m.
How long does it take Clara to get to the library?

B5 Des is baking some biscuits. They take 15 minutes to cook.
If he puts them in the oven at 4:10 p.m, at what time are they ready?

B6 (a) Copy and complete this diagram.

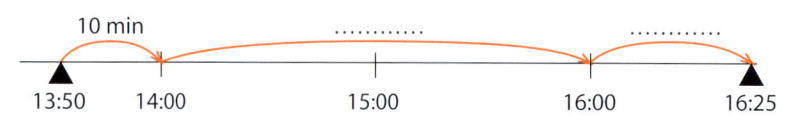

(b) How long is it from 13:50 to 16:25?

B7 How long is it
(a) from 08:15 to 11:30
(b) from 10:45 to 13:15
(c) from 6:15 a.m. to 10:45 a.m.
(d) from 11:30 to 13:15
(e) from 14:40 to 16:25
(f) from 8:45 p.m. to 11:35 p.m.

B8 Randeep gets a train from Birmingham to London.
The train leaves Birmingham at 7:40 a.m.
It takes 1 hour 40 minutes to get to London.
What time does Randeep arrive in London?

c Working out starting times　　　　　　　　　　　　　　level 3

> I need to be at the station at 7:45.
> It takes 20 minutes to walk there.
> When should I leave home?

> The film starts at 8:15.
> It takes 45 minutes to get to the cinema.
> When should I leave home?

C1 What time is it two hours before

(a) 13:00　　　　(b) 5:20 p.m.　　　　(c) half past midday

C2 What time is it 20 minutes before

(a) 8:40 a.m.　　　　(b) 16:55　　　　(c) quarter to ten in the morning

C3 (a) (i) Copy and complete this diagram.

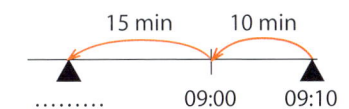

15 min　　10 min
………　　09:00　　09:10

(ii) What time is it 25 minutes before 09:10?

(b) What time is it

(i) 30 minutes before 11:15 a.m.　　(ii) 20 minutes before 20:05

(iii) 45 minutes before 8:20 p.m.　　(iv) 55 minutes before 04:30

C4 Jayden needs to be at work at 8:30 a.m.
It takes him 50 minutes to drive to work.
What time should he leave home?

C5 Olivia is roasting a chicken.
She wants it to be ready at 7:00 p.m.
It will take 2 hours 20 minutes to cook.
What time should she put it in the oven?

C6 Scott is going on holiday.
His flight leaves at 11:30 a.m.

(a) He must check in for the flight at least 2 hours before the flight is due to leave.
What is the latest time he can check in for his flight?

(b) Scott needs to allow $2\frac{1}{2}$ hours to travel from home to the airport.
What is the latest time he can leave home?

D Timetables
level 4

Here is part of a morning bus timetable.

Moor Street	07:20	07:50	08:20
Green Lane	07:30	08:00	08:30
Hobmoor Road	07:35	08:05	08:35
Yew Tree Lane	07:45	08:15	08:45
Garretts Green Lane	07:50	08:20	08:50

When does the 07:20 from Moor Street arrive at Yew Tree Lane?

How long does it take to get from Moor Street to Garretts Green Lane?

How long does it take to get from Hobmoor Road to Yew Tree Lane?

When does the 08:00 from Green Lane get to Hobmoor Road?

- Use a local bus timetable to make up some questions of your own.

Here is the timetable for early morning trains from Norwich to London.

Mondays to Fridays	✗	R ■	✗	R ✗	■	R ■		■ ✗
Norwich	0600	0630	0655	0710	0724	0751	0800	0830
Diss	0618	0647	0713	0727	0741		0817	0847
Stowmarket	0631	0659	0725	0740	0753		0829	0859
Ipswich	0642	0710	0737	0752	0807	0830	0842	0912
Manningtree	0651	0721			0801	0816	0852	0921
Colchester	0703	0732			0812	0827	0902	0931
London	0756	0828	0848	0905	0926	0937	0955	1020

✗ restaurant   ■ buffet/trolley   R reservations recommended

D1 How many of the trains shown stop at Colchester?

D2 What time does the 0724 train from Norwich arrive in London?

D3 How long does the 0600 train from Norwich take to get to London?

D4 (a) How long does the 0751 train from Norwich take to get to London?
(b) How much quicker is this than the 0600 train?

D5 Sarah wants to arrive in London by ten in the morning.
What is the latest train she can get from Diss station?

D6 Jim wants to be in Colchester by 9 a.m.
What is the latest train he can catch from Ipswich?

D7 Mervyn misses the 0655 train from Norwich and waits for the next train.
How much later will he arrive in London than he should have done?

Railroaded

You need sheets FT–12, FT–13 and a dice for this game.

Wick is the most northerly station in the UK with a regular train service. Can you get from Wick to the port of Dover in one day by rail?

Sheet FT–12 has the timetables that will take you from Wick to Dover.

All trains leave on time.
However, the train's arrival might be delayed.

For each train you travel on, roll a dice.   Score 6 and the train arrives 10 minutes early.
Score 5 and the train arrives on time.
Score 1 and the train arrives 10 minutes late.
Score 2 and the train arrives 20 minutes late …

Record your journey on sheet FT–13.

Test yourself

T1 (a) A television programme starts at quarter to eight and lasts for 25 minutes. At what time does it end?

(b) Later, there is a film lasting 1 hour 50 minutes and a quiz lasting 45 minutes. What is the total time for the film and the quiz?

AQA

T2 Teresa spent 40 minutes in the library.
The clock shows the time she left the library.

(a) At what time did she arrive?

(b) After leaving the library she went straight home.
She arrived home at 1112.
How long did it take her to get home?

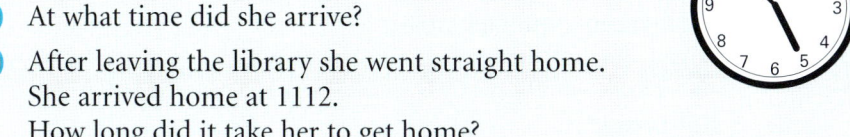

OCR

T3 Here is part of a train timetable from Crewe to London.

(a) At what time should the train leave Coventry?

The train should arrive in London at 1045.

(b) How long should the train take to travel from Crewe to London?

Station	Time of leaving
Crewe	0800
Wolverhampton	0840
Birmingham	0900
Coventry	0930
Rugby	0940
Milton Keynes	1010

Verity arrived at Milton Keynes station at 0953.

(c) How many minutes should she have to wait before the 1010 train leaves?

Edexcel

15 Angle

You will revise
- drawing, measuring and estimating angles
- the names of the types of angles

This work will help you
- deal with angles in special shapes
- use angles for turning, and distinguish between clockwise and anticlockwise

You need an angle measurer.

A Drawing, measuring and sorting angles level 4

If you are unsure about drawing or measuring angles, get your work checked as you go through this section.

A **full turn** is equal to 360°.

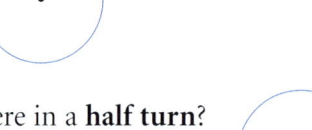

A1 How many degrees are there in a **half turn**?

A2 How many degrees are there in a **quarter turn**?

A quarter turn is called a **right angle**.

A3 Every square has a right angle at each corner and all four sides the same length.

Is this shape a square?
Check by measuring its sides with a ruler and its angles with an angle measurer.

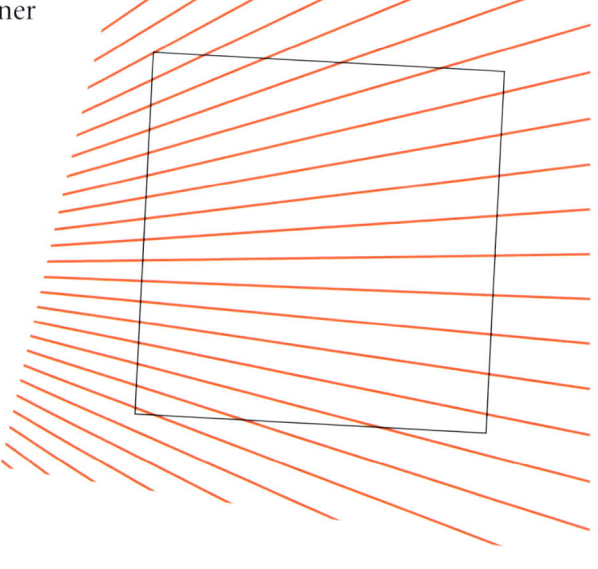

An **acute angle** is one that is less than 90° (less than a right angle).

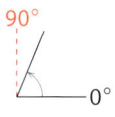

A4 Which of these are acute angles?

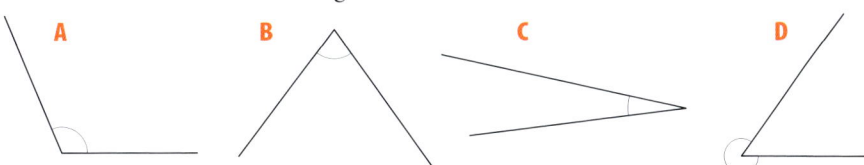

  A   B   C   D

A5 These three angles are acute angles.

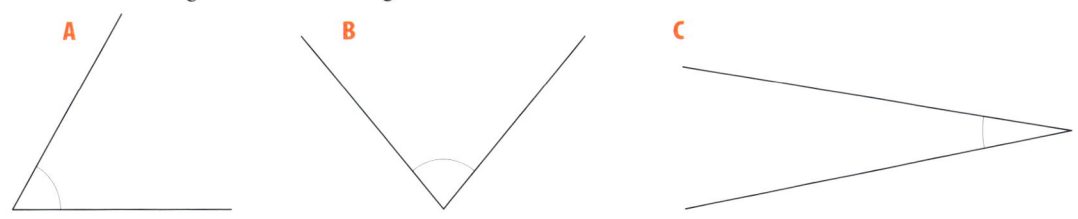

  A   B   C

  (a) Write their letters in order of size, starting with the smallest.
  (b) Measure each angle. Check whether these results agree with your order of size.

A6 Draw these angles.

  (a) 40°   **(b)** 70°   **(c)** 85°   **(d)** 15°

An **obtuse angle** is one that is between 90° and 180° (between a quarter turn and a half turn).

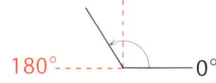

A7 Which of these are obtuse angles?

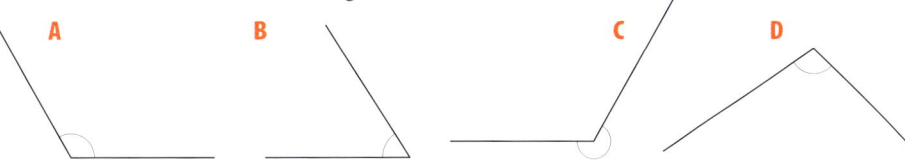

  A   B   C   D

A8 These three angles are obtuse angles.

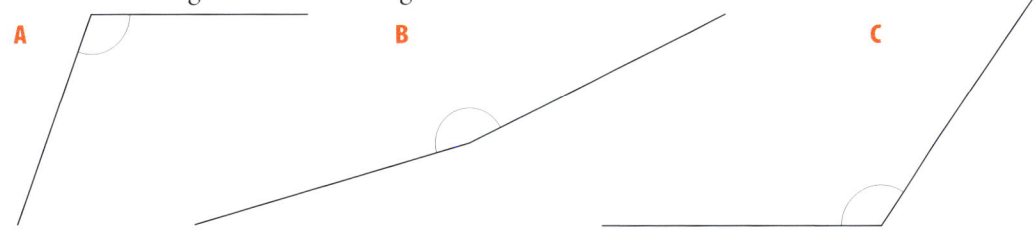

  A   B   C

  (a) Write their letters in order of size, starting with the smallest.
  (b) Measure each angle. Check whether these results agree with your order of size.

A9 Draw these angles.

  (a) 140°   **(b)** 100°   **(c)** 95°   **(d)** 175°

A **reflex angle** is one that is between 180° and 360° (between a half turn and a full turn).

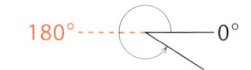

A10 Which of these are reflex angles?

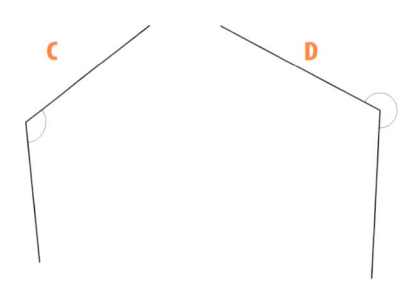

    A   B   C   D

A11 These three angles are reflex angles.

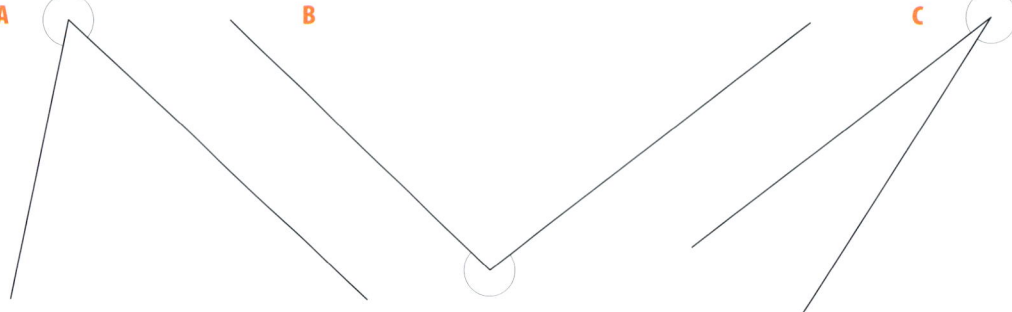

    A   B   C

(a) Write their letters in order of size, starting with the smallest.
(b) Measure each angle. Check whether these results agree with your order of size.

A12 Draw these reflex angles carefully.
(a) 185° (b) 280° (c) 265° (d) 335°

A13 Give the special name (acute, right angle, obtuse or reflex) for each of these angles.
(a) (b) (c) (d)

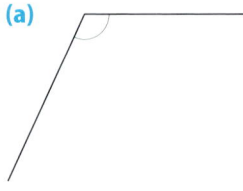

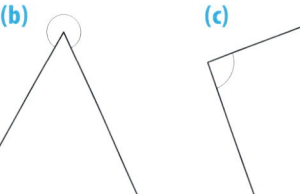

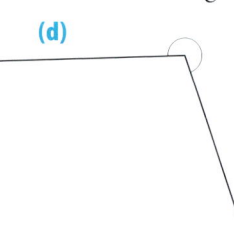

A14 Copy and complete the statements below about the angles in this diagram.
Angle __ is a right angle.
Angle __ is obtuse.
Angle __ is reflex.
Angle __ is acute.

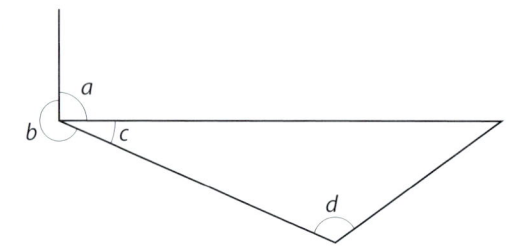

B Turning   level 4

B1 Pam looks at her clock at 9 a.m.
She looks at it again at 12 noon.
What type of angle (acute, right angle, obtuse, half turn or reflex) has the hour hand turned through?

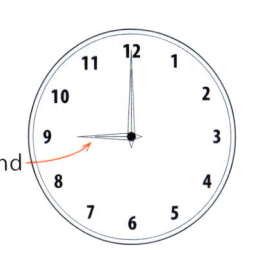

B2 Through what type of angle does the hour hand of a clock turn

(a) between 12 noon and 4 p.m.

(b) between 3 p.m. and 5 p.m.

(c) between 6 a.m. and 1 p.m.

(d) between 5 p.m. and 7 p.m.

(e) between 2 p.m. and 8 p.m.

You could make up some more questions like those in B2 and try them on a partner.

The pointer on this circle can move clockwise ↷ or anticlockwise ↶.

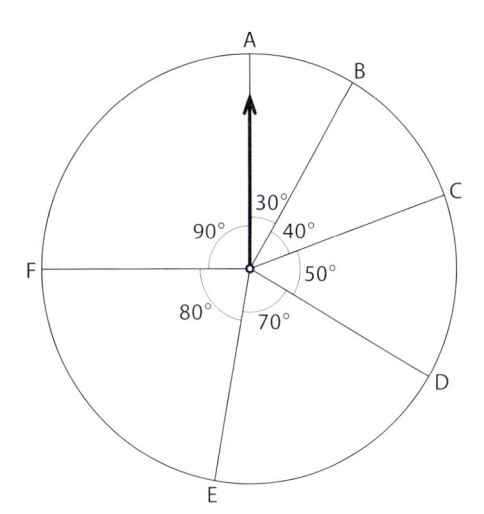

B3 Zoe turns the pointer clockwise from A to C.

(a) What angle has it turned through?

(b) What type of angle is it (acute, right angle, obtuse, half turn or reflex)?

B4 For each of these moves, give the angle the pointer turns through and the type of angle it is.

(a) Clockwise from C to E

(b) Anticlockwise from D to B

(c) Clockwise from A to E

(d) Anticlockwise from B to F

B5 Tom turns the pointer through a reflex angle from F to C.
Did he turn it clockwise or anticlockwise?

C Angles in shapes

C1 (a) Measure the sides of this triangle accurately.
What special kind of triangle is it?

(b) Measure the angles.
What do you notice about the angles?

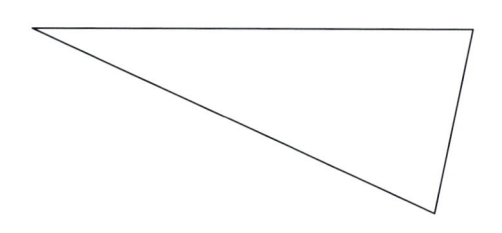

C2 (a) Measure the sides of this triangle accurately.
What special kind of triangle is it?

(b) Measure the angles.
What do you notice about them?

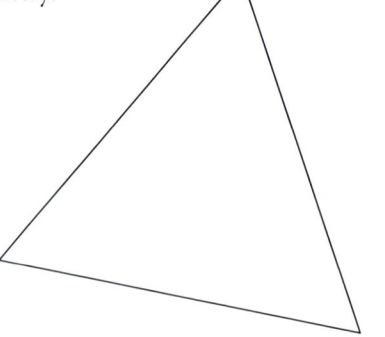

C3 (a) On this shape, measure the sides and angles carefully and say which of them are equal.

(b) Give the special name of the shape.

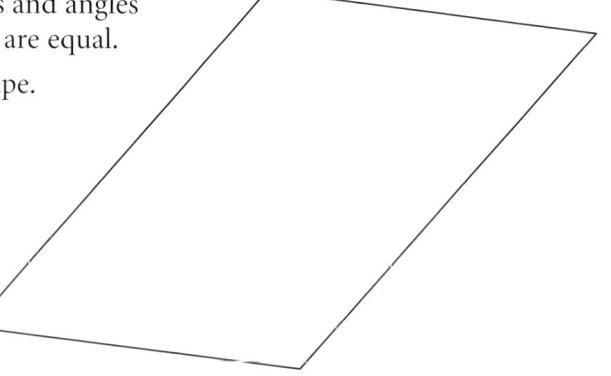

A triangle with a right angle in it is called a **right-angled triangle**.

C4 Here is a sketch of a right-angled triangle.

(a) Draw the triangle accurately.

(b) Measure and record
   (i) angle X
   (ii) angle Y
   (iii) the length of side XY

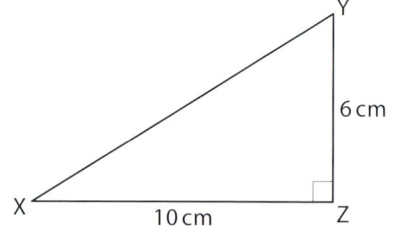

C5 This sketch shows another right-angled triangle.

(a) Draw the triangle accurately.

(b) Measure and record the length of side QR and side PR.

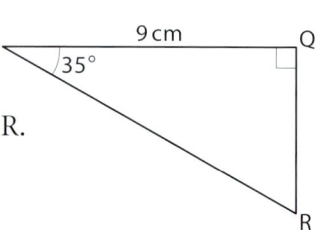

D Estimating angles

In this section, do not measure the angles.

D1 Here is an angle of 45° (half of 90°).
Say whether each of the acute angles below
is less than 45° or more than 45°.

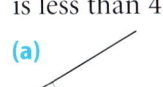

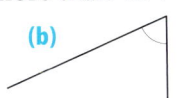

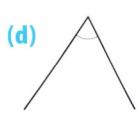

(a) (b) (c) (d)

D2 Write an estimate, to the nearest 10°, of each angle in question D1.

D3 This reflex angle is 270° (halfway between 180° and 360°).
Say whether each of the reflex angles below
is less than 270° or more than 270°.

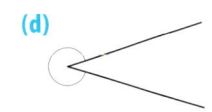

(a) (b) (c) (d)

D4 Write an estimate, to the nearest 10°, of each angle in question D3.

D5 This obtuse angle is 135° (halfway between 90° and 180°).
Say whether each of the obtuse angles below
is less than 135° or more than 135°.

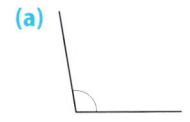

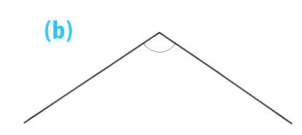

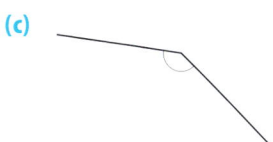

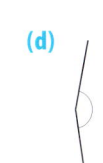

(a) (b) (c) (d)

D6 Write an estimate, to the nearest 10°, of each angle in question D5.

D7 Holly draws this angle.
Which of the angles below is
closest to **twice** Holly's angle?

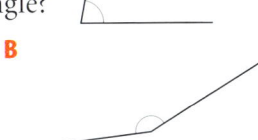

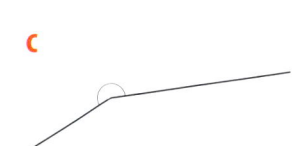

A B C

D8 Naomi draws this angle.
Which of the angles below is
closest to **half** Naomi's angle?

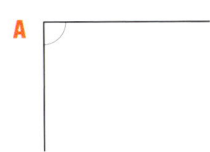

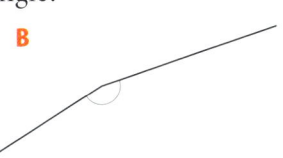

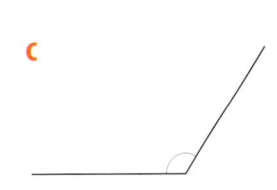

A B C

Test yourself

T1 (a) Measure and write down the size of the angle *p*.
(b) Write down what type of angle it is.

AQA

T2 Measure the obtuse angle in this triangle.

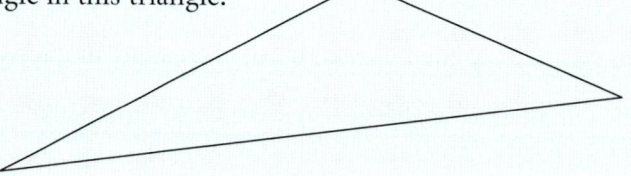

T3 Lydie faces north.
She turns clockwise to north-east.
(a) What angle has she turned through?
(b) What type of angle is it (acute, right angle, obtuse, half turn or reflex)?

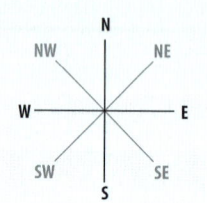

T4 Lydie does some more turns. For each of them, give the size of the angle she turns through and the type of angle it is.
(a) Clockwise from facing east to facing south-west
(b) Anticlockwise from facing south to facing west
(c) Clockwise from facing west to facing south-east
(d) Anticlockwise from facing south-west to facing north-east

T5 Lydie turns through an obtuse angle from north-west to south. Does she turn clockwise or anticlockwise?

T6 This is a sketch of a triangle.
(a) Draw the triangle accurately.
(b) Measure the length BC.

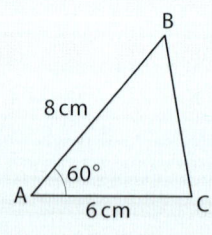

T7 Write an estimate of each of these angles, to the nearest 10°.
(a)   (b)   (c)

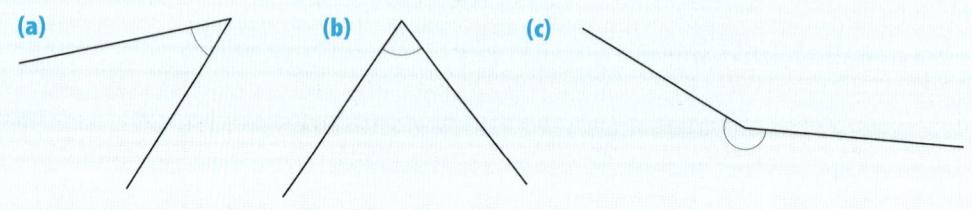

16 Length

This work will help you use millimetres, centimetres, metres and kilometres.

A Centimetres and millimetres

Building materials such as screws and nails usually have their lengths given in millimetres (mm).

This nail is 45 mm or 4.5 cm long.

- How long is a 23 mm nail in cm?
- What is 7.8 cm in mm?

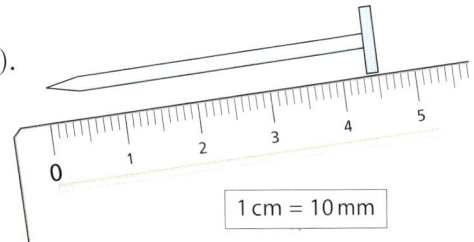

1 cm = 10 mm

A1 Measure the lengths of these nails. Write down their measurements in   (i) mm   (ii) cm

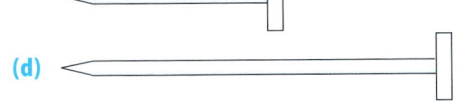

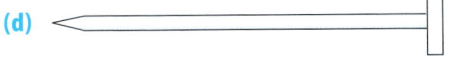

(a)   (b)   (c)

(d)   (e)

(f)

A2 Use a ruler to draw a straight line that has a length of 56 mm.

A3 Draw a rectangle that has a length of 6.4 cm and a width of 3.8 cm.

A4 Write 110 mm in cm.

A5 What completes each statement, mm or cm?

(a) My thumbnail is 15 ___ long.   (b) My pencil is 6 ___ long.

(c) My son's height is 168 ___ .   (d) I have a paperclip that is 7 ___ wide.

A6 Spiders are often measured by how wide their legs spread.
Copy and complete this table showing the leg-span of some spiders.

Spider	Leg-span (mm)	Leg-span (cm)
Bird-eating Spider	250 mm	
Tarantula		24 cm
Raft Spider (UK)	145 mm	
House Spider (UK)		7.5 cm
Wolf Spider (UK)	17 mm	
Money Spider (UK)	3 mm	

16 Length 75

A7 Put these lengths in order, shortest first.

5 cm   5 mm   2.7 cm   33 mm $2\frac{1}{2}$ cm

A8 Eve was 161.7 cm tall on her fifteenth birthday.
In the year that followed she grew 6 mm.
How tall was she on her sixteenth birthday?

B Metres and centimetres

1 m = 100 cm

- How many centimetres are in 3 metres? What about 3.25 m?
- How would you convert 520 cm to metres?

B1 A curtain has a width of 4 metres.
What is this width in centimetres?

B2 How many cm are in each of these?

(a) 2 m   (b) 10 m   (c) $4\frac{1}{2}$ m   (d) $\frac{1}{4}$ m

B3 What completes each statement, cm or m?

(a) My hand is 17 __ long.
(b) My young daughter is 1 __ tall.
(c) My foot is 8.3 __ wide.
(d) My car is 4.5 __ long.

B4 Write the lengths of these snakes in centimetres.

Python 6 m   Anaconda 8.5 m   Grass Snake 1.25 m   Adder 0.75 m

B5 The width of a window is 200 cm.
What is this width in metres?

B6 Write the wingspans of these bats in metres.

Fruit Bat 170 cm   Mouse-eared Bat (UK) 45 cm
Pipistrelle Bat (UK) 25 cm   Kitti's Hog-nosed Bat 9 cm

B7 David buys a plank of wood that is 1.8 m long.
He cuts off 40 cm.
How long is the piece that is left?

B8 Dan was 1.18 m tall on his birthday.
In the year that followed he grew 7 cm.
How tall was he on his next birthday?

B9 Put each set of lengths in order, shortest first.

(a) 30 m, 3 cm, 30 cm, 3 m
(b) 76 cm, 2.3 m, 150 cm, 1.36 m
(c) 15 cm, 1.2 m, $1\frac{1}{4}$ m, 105 cm
(d) 5 mm, 5 m, 50 cm, 50 mm

C Kilometres

A kilometre is a thousand metres.

Make a list of places in and around your school that you think are
- more than 1 kilometre away from your classroom
- less than 1 kilometre away from your classroom

Check with a map.

C1 How many metres are there in each of these?
   (a) 5 km   (b) 2 km   (c) $\frac{1}{2}$ km   (d) $\frac{3}{4}$ km

C2 How many kilometres is each of these?
   (a) 4000 m   (b) 3000 m   (c) 20 000 m   (d) 100 000 m

C3 How many metres are there in each of these?
   (a) 2.5 km   (b) 9.4 km   (c) 0.6 km   (d) 0.25 km

C4 How many kilometres is each of these?
   (a) 1500 m   (b) 5800 m   (c) 300 m   (d) 2640 m

C5 Rahima was running in a 15 km road race.
She dropped out only 800 metres from the end.
How far had she run?

Test yourself

T1 (a) Measure this line. ⎯⎯⎯⎯⎯⎯⎯⎯⎯⎯⎯⎯
   Give your answer in centimetres.
   (b) (i) Draw a straight line that is 7 cm long.
   (ii) Find the point that is halfway along the line you have drawn.
   Mark it with a cross (×).

T2 (a) A pencil is 72 mm long. How many centimetres is 72 mm?
   (b) A lawn is 5.4 m wide. How many centimetres is 5.4 m?

T3 Put these four lengths in order, shortest first.
   2 mm   2 m   20 cm   20 mm
   OCR

T4 A path is 5.6 km long.
Jane walks 4.9 km along this path.
How many metres does she need to walk to get to the end of the path?

17 Squares and square roots

This work will help you

- use square numbers and their square roots
- use the notation 6^2 for square numbers and the notation $\sqrt{9}$ for square roots
- use a calculator to find squares and square roots

A Square numbers and square roots   level 4

Leo makes 16 buns.
He arranges them in rows to make a square.

If a number of things can be arranged in a square like this, then the number is a **square number**.

So 16 is a square number.

- Which of the numbers below is a square number?

   3   18   19   20   24   25   30   40

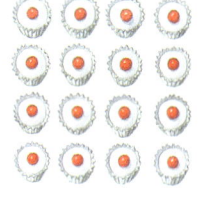

A1 Show how to arrange 9 counters in rows to make a square.

A2 (a) Is 36 is a square number?

(b) Which of the numbers below is a square number?

   2   4   6   8   10   12   14

A3 Leo wants to decorate a cake by arranging chocolate stars in rows to make a square. He has 50 chocolate stars and he makes the biggest square he can with them.

(a) How many stars does he use altogether?

(b) How many rows of stars are on his cake?

(c) How many stars are left over?

A4 Use the clues to find the numbers.

(a) A square number
Less than 10
An even number

(b) A square number
Less than 100
A multiple of 5

(c) A square number
Less than 100
More than 70

A5 List all the square numbers that are smaller than 150.

A6 121 counters are arranged in a square.
How many rows of counters are there?

---

16 is a square number as we can arrange 16 counters in 4 rows of 4 counters.

We say that the **square root** of 16 is 4.

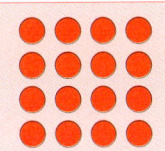

A7 Find the square root of each of these numbers.

(a) 9 (b) 25 (c) 100 (d) 36 (e) 64

B Using shorthand

When you multiply any whole number by itself you get a square number. Multiplying a number by itself is called **squaring**.

16 is a square number because $4 \times 4 = 16$.
We say that 4 **squared** is 16.

Using shorthand, we can write this as $4^2 = 16$.

B1 Find the value of each of these.

(a) 3 squared (b) 10 squared (c) 8 squared

B2 Copy and complete: $10^2 = 10 \times 10 = \blacksquare$.

B3 Sort these into four matching pairs.

| 3^2 | 7^2 | 1^2 | 8^2 | 1 | 64 | 9 | 49 |

B4 Find the value of each of these.

(a) 5^2 (b) 6^2 (c) 2^2 (d) 9^2 (e) 12^2

$4 \times 4 = 16$
So the square root of 16 is 4.
Using shorthand, we can write this as $\sqrt{16} = 4$.

B5 Find the value of each of these.

(a) $\sqrt{25}$ (b) $\sqrt{4}$ (c) $\sqrt{1}$ (d) $\sqrt{81}$ (e) $\sqrt{100}$

B6 What is the missing number in each of these?

(a) $\blacksquare$ squared is 9 (b) The square root of $\blacksquare$ is 6
(c) $\blacksquare^2 = 25$ (d) $\sqrt{\blacksquare} = 4$

C Using a calculator

A calculator has a key for squaring numbers.
- Find this key on your calculator and use it to work these out.

  A 3^2 **B** 10^2 **C** 12 squared

A calculator also has a key for finding square roots.
- Find this key on your calculator and use it to work these out.

  A $\sqrt{121}$ **B** $\sqrt{81}$ **C** the square root of 196

C1 Use your calculator to find the value of each of these.
(a) 5 squared
(b) 11 squared
(c) 19 squared
(d) 4^2
(e) 13^2
(f) 25^2

C2 Use your calculator to find the value of each of these.
(a) the square root of 64
(b) the square root of 196
(c) $\sqrt{49}$
(d) $\sqrt{324}$
(e) $\sqrt{400}$
(f) $\sqrt{961}$

C3 A gardener likes to plant rose bushes in rows to make square designs.
He plants 144 rose bushes like this.
How many rows of bushes does he plant?

C4 Find all the square numbers between 200 and 300.

C5 Rearrange these cards to make a square number. 2 9 7

Test yourself

T1 Which of the numbers in the box are square numbers?   1   25   40   4   2   9   81   80

T2 Find the value of each of these.
(a) 9 squared
(b) 5^2
(c) 8^2
(d) the square root of 49
(e) $\sqrt{9}$
(f) $\sqrt{36}$

T3 Use a calculator to find the value of each of these.
(a) 17 squared
(b) 21^2
(c) 30^2
(d) the square root of 169
(e) $\sqrt{256}$
(f) $\sqrt{10\,000}$

T4 What is the smallest square number that is bigger than a thousand?

18 Adding and subtracting decimals

This work will help you add and subtract decimals without using a calculator.

You need sheet FT–14.

A Adding decimals
level 4

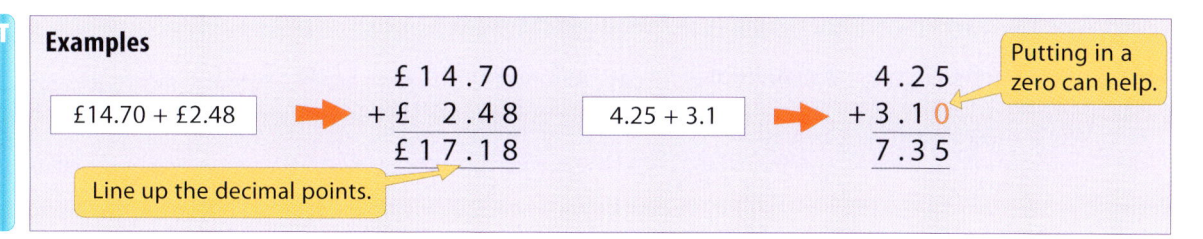

Examples

£14.70 + £2.48 ➡
```
  £ 1 4 . 7 0
+ £     2 . 4 8
  £ 1 7 . 1 8
```
Line up the decimal points.

4.25 + 3.1 ➡
```
  4 . 2 5
+ 3 . 1 0
  7 . 3 5
```
Putting in a zero can help.

A1 Work these out in your head.
(a) 0.3 + 0.5   (b) 1 + 6.5   (c) 7 + 0.3   (d) 2.4 + 3
(e) 7.9 + 2   (f) 3.2 + 0.4   (g) 0.2 + 1.3   (h) 2.5 + 3.5

A2 Work these out in your head.
(a) £1 + £0.40   (b) £0.60 + £3   (c) £1.20 + £2.50   (d) £15 + £1.20

A3 Hayley buys a magazine that costs £2.75 and a can of drink that costs £0.85.
How much does she spend altogether?

A4 Work these out.
(a) £3.42 + £1.05   (b) £5.63 + £1.19   (c) £24.76 + £5.62   (d) £15.89 + £36.16

A5 Jim joins two pieces of wood end to end.
One piece is 1.34 m long and the other is 2.68 m long.
What is the total length of these pieces?

A6 Work these out.
(a) 1.5 m + 4.7 m   (b) 2.36 m + 1.45 m   (c) 4.9 m + 15.74 m

A7 Work these out.
(a) 4.5 + 3.9   (b) 8.3 + 3.8   (c) 12.4 + 7.6   (d) 23.5 + 91.7

A8 Find two numbers in the loop that add to give
(a) 2.5   (b) 1.6   (c) 1.86
(d) 2.64   (e) 2.86   (f) 3

(1.56   1.3   1.44   1.2   0.3)

A9 Work these out.
(a) 1.47 + 3.28   (b) 2.46 + 10.91   (c) 16.04 + 9.5   (d) 30.85 + 9.4

B Subtracting decimals

level 4

Example

5.34 − 1.5

$$\begin{array}{r} 5.34 \\ -1.50 \\ \hline \end{array}$$

Putting in a zero can help.

$$\begin{array}{r} ^4\!5.^13\,4 \\ -1.5\,0 \\ \hline 3.8\,4 \end{array}$$

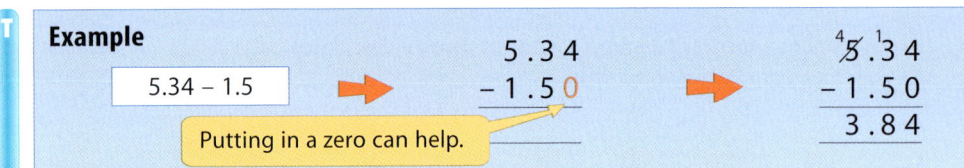

B1 Work these out in your head.
(a) 3.2 − 0.2 (b) 6.4 − 0.1 (c) 4.9 − 0.5 (d) 6.4 − 1
(e) 4.9 − 3   (f) 8.3 − 8   (g) 8 − 0.5   (h) 1 − 0.2

B2 Work these out in your head.
(a) £1 − £0.20 (b) £4 − £1.50 (c) £1.50 − £0.40 (d) £2.50 − £0.90

B3 Work these out.
(a) £6.95 − £3.25 (b) £4.37 − £1.19 (c) £12.08 − £3.40 (d) £45.21 − £44.90

B4 Asif has a piece of wood that is 3.6 metres long. He cuts off 1.5 metres.
What is the length of the remaining piece?

B5 Yasmin has £12.55 in her purse. She buys a top that costs £5.99.
How much money does she have left?

B6 Work these out.
(a) 16.2 − 3.1   (b) 17.89 − 3.06 (c) 6.7 − 4.9   (d) 2.65 − 1.28
(e) 4.21 − 2.38 (f) 5.29 − 4.1   (g) 8.56 − 3.7 (h) 21.61 − 13.9

B7 Sue is knitting the back of a jumper that should measure 25 cm.
She has knitted 11.4 cm so far.
How much has she left to knit?

B8 A pole is painted black and blue.
What is the length of the blue part?
(2.6 m total; 1.43 m black)

Example

Here, putting in a zero helps a lot.

8.5 − 3.14

$$\begin{array}{r} 8.5\,0 \\ -3.1\,4 \\ \hline \end{array}$$

$$\begin{array}{r} 8.^4\!5^1\!0 \\ -3.1\,4 \\ \hline 5.3\,6 \end{array}$$

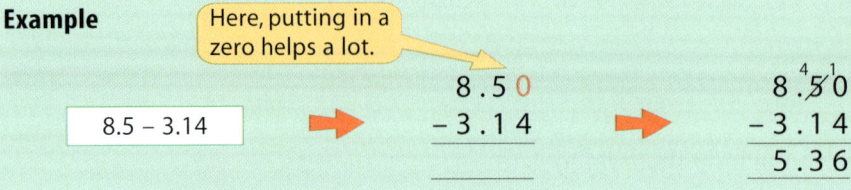

B9 Work these out.
(a) 4.6 − 2.23   (b) 8.3 − 6.12   (c) 5.4 − 1.39
(d) 12.4 − 0.65 (e) 51.2 − 14.36   (f) 3.8 − 1.56

B10 Calculate £5 − £1.26.

B11 Calculate these. (a) 6 − 3.12   (b) 7 − 5.83   (c) 4 − 3.71

C Mixed questions
level 4

C1 Work these out.

(a) £1.65 + £3.15   (b) £5.45 − £1.30   (c) £6.13 + £2.95   (d) £3 − £1.25

C2 Six guesses for the height of a tree are shown.

A 5.13 m   B 5.3 m   C 4.9 m   D 4.7 m   E 4.74 m   F 5 m

The real height of the tree is 4.85 metres.

(a) What is the difference between each guess and the real height?
(b) Which guess is closest?

C3 The perimeter of this triangle is 8 m.
Find the length of the longest edge.

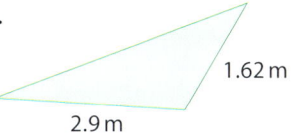

1.62 m
2.9 m

> The perimeter is the total distance round the edges.

C4 Do the puzzles on sheet FT–14.

Test yourself

T1 (a) Suneet buys two books. The books are priced at £4.95 and £5.75.
How much does he spend altogether on these books?

(b) Pia has £4.60 in her purse.
She spends £1.75 on a coffee and a bun.
How much money has she left?

T2 (a) What is the total weight of these cats?
(b) What is the difference between the weights?

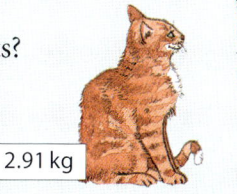

3.58 kg
2.91 kg

T3 A pole is painted orange and white.
What is the length of the orange part?

4 m
2.35 m

T4 Work these out.

(a) 3.5 − 2.8   (b) 7.12 + 4.73   (c) 9 − 2.3   (d) 7.7 − 4.26

18 Adding and subtracting decimals

19 Mental methods 3

This work will help you with some ways of multiplying and dividing mentally by 4 and 5.

A Multiplying and dividing by 4 and 5 *level 4*

4 is 2×2.

To **multiply by 4** you can multiply by 2 and then by 2 again.

14 × 4   [14] →(×2)→ [28] →(×2)→ [56]

To **divide by 4** you can divide by 2 and then by 2 again.

52 ÷ 4   [52] →(÷2)→ [26] →(÷2)→ [13]

Do these in your head, by any method you like.

A1 (a) 13×4 (b) 23×4 (c) 56÷4 (d) 4×25 (e) 4×32
(f) 4×18 (g) 92÷4 (h) 4×34 (i) 108÷4 (j) 420÷4

5 is 10÷2.

To **multiply by 5** you can multiply by 10 and then divide by 2 (or the other way round).

43 × 5   [43] →(×10)→ [430] →(÷2)→ [215]   38 × 5   [38] →(÷2)→ [19] →(×10)→ [190]

To **divide by 5** you can divide by 10 and then multiply by 2 (in either order).

160 ÷ 5   [160] →(÷10)→ [16] →(×2)→ [32]   315 ÷ 5   [315] →(×2)→ [630] →(÷10)→ [63]

A2 (a) 24×5 (b) 28×5 (c) 34×5 (d) 72×5 (e) 5×66
(f) 5×18 (g) 5×35 (h) 63×5 (i) 26×5 (j) 5×23

A3 (a) 340÷5 (b) 480÷5 (c) 130÷5 (d) 420÷5 (e) 1620÷5
(f) 660÷5 (g) 520÷5 (h) 135÷5 (i) 215÷5 (j) 1040÷5

A4 (a) 130×4 (b) 5×48 (c) 170÷5 (d) 164÷4 (e) 26×5
(f) 4×98 (g) 720÷5 (h) 102×4 (i) 5×81 (j) 220÷4

Review 3

1. Do these in your head.
   - (a) 27×4
   - (b) 5×72
   - (c) $308 \div 4$
   - (d) $2 + 3.4$
   - (e) $7 - 0.5$

2. Write the following times in 12-hour clock time, using a.m. or p.m.
   - (a) A quarter past eight in the morning
   - (b) 03:10
   - (c) 16:25

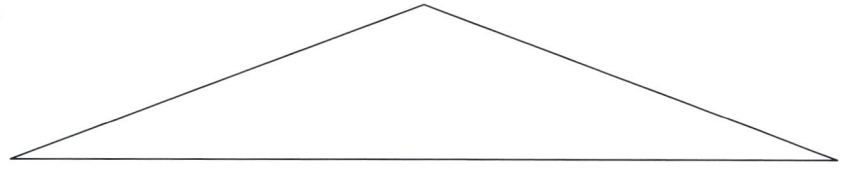

3.
   - (a) Measure each side of this triangle in millimetres.
   - (b) What kind of triangle is it?
   - (c) (i) Is the largest angle in this triangle acute, obtuse or reflex?

     (ii) Measure the size of this angle.

4. A circle has a diameter of 9 cm.
   Write down the length of its radius
   - (a) in cm
   - (b) in mm

5. What is the value of
   - (a) the square root of 9
   - (b) $\sqrt{100}$

6. Work these out.
   - (a) $1.9 + 3.6$
   - (b) $4.75 - 1.32$
   - (c) $14.52 + 2.6$
   - (d) $12.6 - 7.19$

7. Kay has a silk ribbon that is 2.6 metres long.
   She cuts off a piece 80 cm long.
   How long is the piece that is left?

8. Here is part of a bus timetable.
   - (a) Gregor needs to be at the railway station by 2:05 p.m. Will this bus from the Parade get him there on time?
   - (b) How long is it between the times for the railway station and Ladymead Road?

Stop	Departure time
Parade	1345
Railway station	1349
Wellsprings Road	1352
Ladymead Road	1355
Priorswood shops	1357

9. Put this set of lengths in order, shortest first.

   5 mm   50 cm   5 m   50 mm   0.5 km

10. Which of the numbers in the loop are
    - (a) square numbers
    - (b) multiples of 5

    (16, 25, 52, 40, 49, 15)

20 Number patterns

This work will help you find missing numbers in number patterns, explaining your methods.

You may need sheets FT–15 and FT–16.

A Simple patterns level 4

Some squares are shaded blue to make a pattern of diagonal lines on this number grid.

The numbers on the blue squares follow the pattern 1, 4, 7, 10, …

- List the first twenty numbers in this pattern.
- The number 100 is on a blue square.
  What is the next number after this on a blue square?
- Which of the numbers below are on blue squares?

  106   60   90   61   59   150   301   85

1	2	3	4	5	6	7	8
9	10	11	12	13	14	15	16
17	18	19	20	21	22	23	24
25	26	27	28	29	30	31	32
33	34	35	36	37	38	39	40
41	42	43	44	45	46	47	48

You could investigate other patterns on this grid. There are some unshaded grids on sheet FT–15.

A1 Here is a number pattern.

   1   7   13   19   25   31   …

(a) What is the next number in the pattern?

(b) Explain how you worked out your answer.

A2 Here is a number pattern.

   40   37   34   31   28   25   22   …

(a) What are the next two numbers in the pattern?

(b) Explain how you worked this out.

A3 Find the next two numbers in each number pattern.

(a) 5, 7, 9, 11, 13, …
(b) 10, 14, 18, 22, 26, …
(c) 26, 29, 32, 35, 38, …
(d) 30, 37, 44, 51, 58, …
(e) 17, 15, 13, 11, 9, …
(f) 30, 26, 22, 18, 14, …

A4 Here is a number pattern.

   6   11   16   21   26   31   …

(a) What is the next number in the pattern?

(b) Write down the first number in this pattern that is bigger than 50.

(c) Which of the numbers below are in the pattern?

   41   63   36   54   81   50   101

A5 Here are six number cards.

| 37 | 53 | 29 | 45 | 13 | 21 |

(a) How could you arrange them to make a number pattern?

(b) Explain how to find the next number in your pattern.

A6 Kayla wants to show the numbers 1 to 100 on a long strip of paper. She colours in the strip like this.

| 1 | 2 | 3 | 4 | 5 | 6 | 7 | 8 | 9 | 10 | 11 | 12 | 13 |

(a) What number is on the next blue square?

(b) (i) Write down the first six numbers that are on grey squares.

(ii) What number is on the 7th grey square?

(iii) Work out which number is on the 20th grey square.

(c) Write down the first six numbers that are on orange squares.

(d) What colour square is each of these on?

(i) 16 (ii) 20 (iii) 21 (iv) 39 (v) 30

B Further patterns

A set of numbers in order is often called a **sequence**.

The squares in the first column of this number grid are all shaded blue.

The numbers on the blue squares make the sequence 1, 2, 4, 7, 11, …

- List the first ten numbers in this sequence.
- Explain how you would find the 11th number in this sequence.

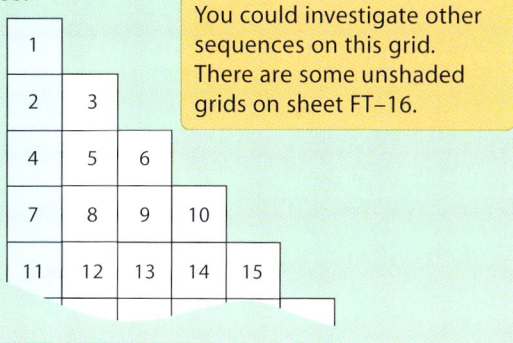

You could investigate other sequences on this grid. There are some unshaded grids on sheet FT–16.

B1 Here is a sequence.

5 6 8 11 15 20 …

(a) What is the next number in the sequence?

(b) Explain how you worked out your answer.

B2 Find the next two numbers in each sequence.

(a) 1, 2, 4, 8, 16, …   (b) 64, 32, 16, 8, 4, …
(c) 6, 7, 9, 12, 16, …   (d) 40, 39, 37, 34, 30 …
(e) 6, 7, 10, 15, 22, …   (f) 6, 7, 9, 13, 21, …

B3 Look at this pattern.
$$1 + 2 + 3 = 6$$
$$2 + 3 + 4 = 9$$
$$3 + 4 + 5 = 12$$
$$\vdots$$

(a) Copy this pattern and show the next two lines.

(b) Copy and complete these so they are lines in the same pattern.

(i) $10 + \ldots + \ldots = \ldots$

(ii) $\ldots + 12 + \ldots = \ldots$

(iii) $\ldots + \ldots + 20 = \ldots$

(iv) $\ldots + \ldots + \ldots = 24$

(c) Explain why $\ldots + \ldots + \ldots = 28$ cannot be completed as a line in this pattern.

B4 (a) Without using a calculator, work out 222×5.

(b) Copy and complete the first five lines of this number pattern.
$$2 \times 5 = 10$$
$$22 \times 5 = 110$$
$$222 \times 5 = \ldots\ldots$$
$$2222 \times 5 = 11\,110$$
$$\ldots\ldots \times \ldots = \ldots\ldots$$

(c) Leonie thinks that $2\,222\,222 \times 5 = 1\,111\,110$ is a line from the same pattern. Show that she is wrong.

Test yourself

T1 (a) Write down the next number in this sequence.

2   6   10   14

(b) Explain how you got your answer.

OCR

T2 Look at this number pattern.

23   28   33   38   43   48   …

(a) (i) What is the next number in the pattern?

(ii) Explain how you worked out your answer.

(b) Explain why 200 is **not** in this number pattern.

T3 Here are five number cards.

| 35 | 31 | 23 | 39 | 27 |

(a) Arrange these cards to make a number pattern.

(b) Explain how to find the next number in your number pattern.

OCR

T4 Find the next two numbers in this sequence.   3,   6,   12,   24,   …

21 Estimating and scales

This work will help you
- estimate lengths by comparison
- read a variety of scales, including decimal scales

You need sheet FT–17.

A Estimating lengths

Imagine a metre rule. Use this image to estimate
- the height of your classroom
- the height of your teacher

A standard ruler is about 30 cm long. Estimate
- the length of your desk
- the height of this book

Lucy is standing next to life-size cut-outs of some of the smallest and tallest people to have lived in the last century.

Lucy is 150 cm tall.
Estimate the heights of the cut-outs.

150 cm

21 Estimating and scales

A1 This pencil is 8 cm long.
Estimate the lengths of these pencils.

8 cm

(a)

(b)

(c)

(d)

(e)

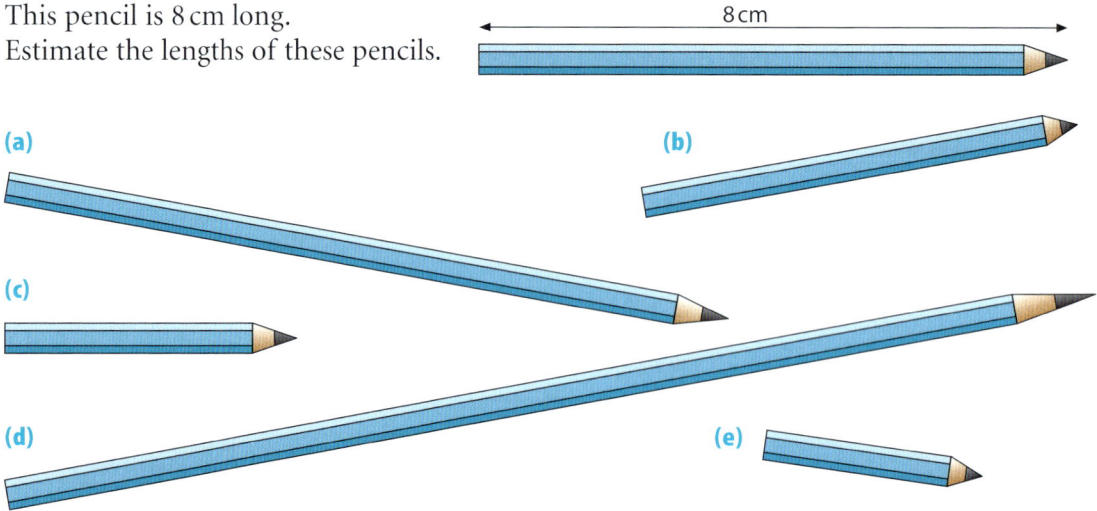

A2 This rectangle is 2 cm high.

(a) Estimate how long the rectangle is.

2 cm

Estimate the length and height of these rectangles.

(b)

(c)

A3 The car in these pictures is about 4 m long.
Estimate the lengths of the other vehicles in these pictures.

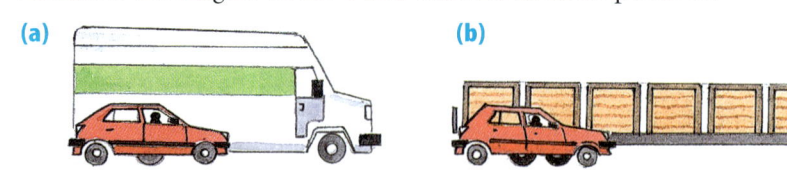

(a)

(b)

(c)

(d)

A4 Danny, who is about 150 cm tall, is standing in the garden.

Estimate

(a) the height of the shed

(b) the height of the bike (to the top of the saddle)

(c) the diameter of the bike's wheels

A5 This is a picture of the Bargate. Two policeman are standing by it.

(a) Estimate the height of one of the policemen. Use this to estimate the height of the Bargate.

(b) Explain how you worked out your estimate.

OCR

A6 The average height of a 16-year-old girl is about 1.6 metres. In metres, estimate the height of

(a) an ordinary door

(b) an average 16-year-old boy

B Reading and estimating from scales

• What readings are shown on these scales?
Remember to include the units each time.

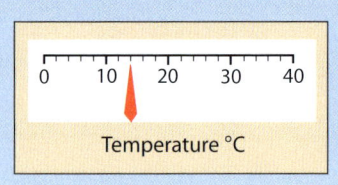

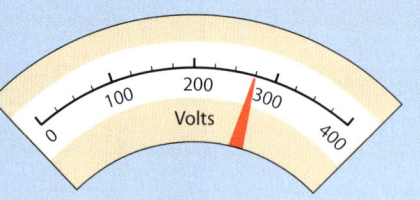

Estimate this reading

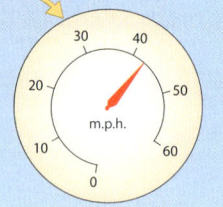

B1 What is the temperature on each of these thermometers?

(a)　　　(b) 　　　(c)

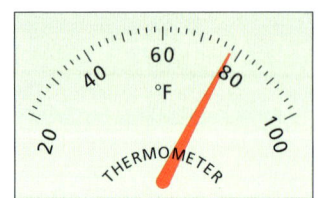

B2 The thermometers below are from an oven.

(a) What does one small division on the scale represent?

(b) What temperature does each of these thermometers show?

(i)　　　(ii) 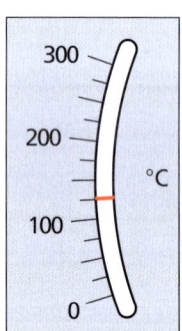　　　(iii)

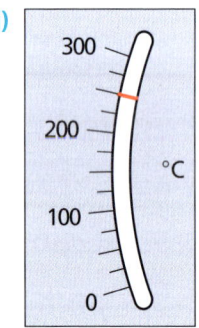

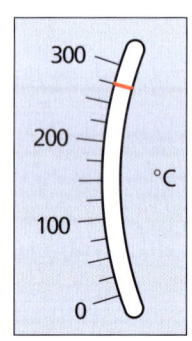

B3 For each of these speedometers

(i) say what one small division represents　　　(ii) give the speed that it shows

(a)　　　(b) 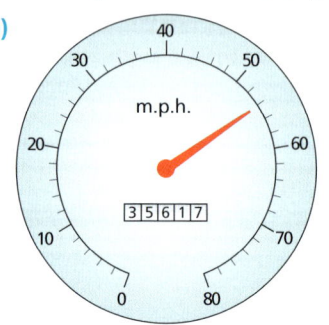　　　(c)

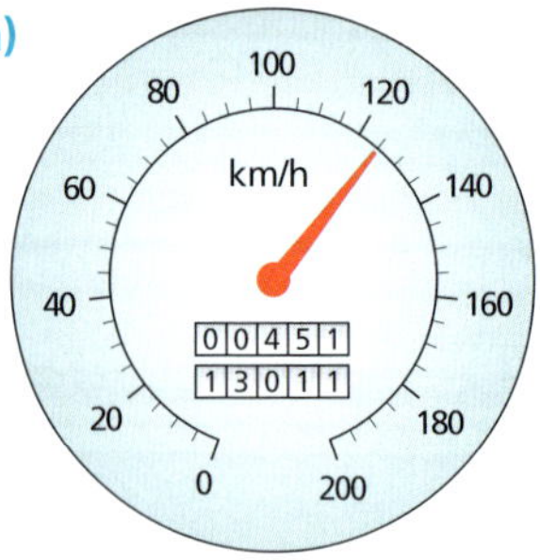

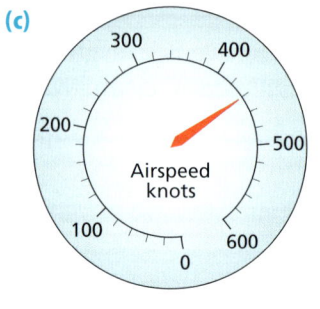

92　21 Estimating and scales

B4 For each of the scales on sheet FT–17 A, draw an arrow to show the reading given.

B5 These speedometers have no small divisions marked.
Estimate the speeds they are showing.

(a)   (b)   (c)

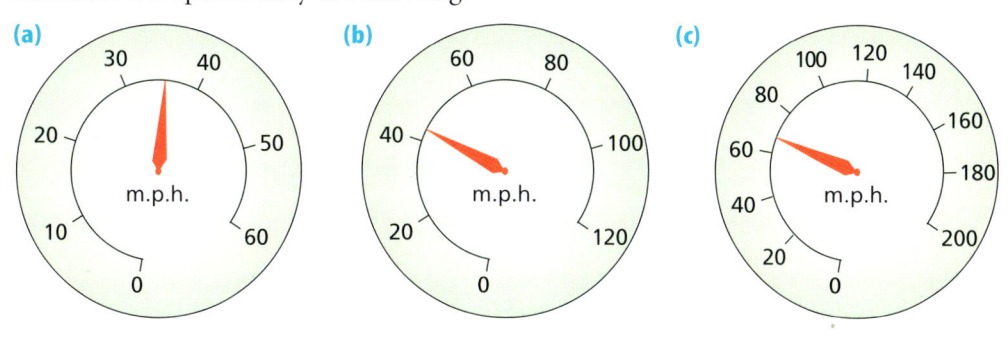

C Decimal scales

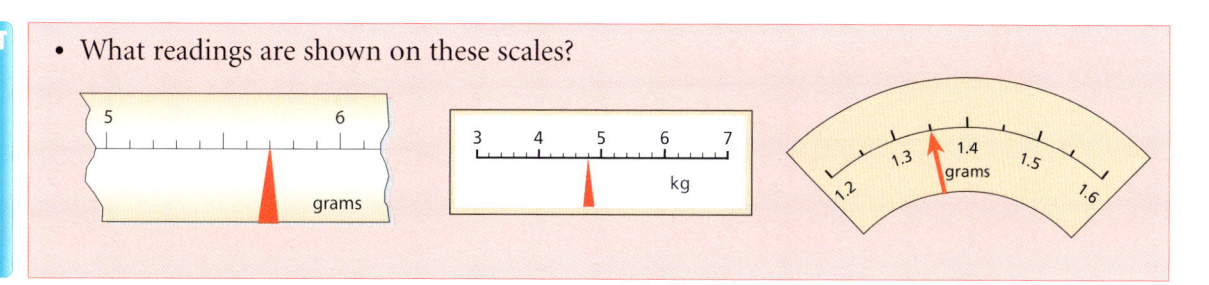

- What readings are shown on these scales?

C1 Joel says this arrow points to 3.2 but Josie says it points to 3.4.
Who is right?

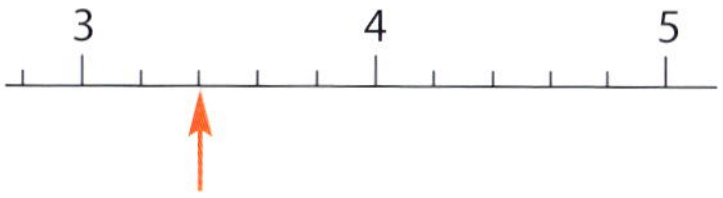

C2 What are the readings on these scales?

(a)   (b)   (c)

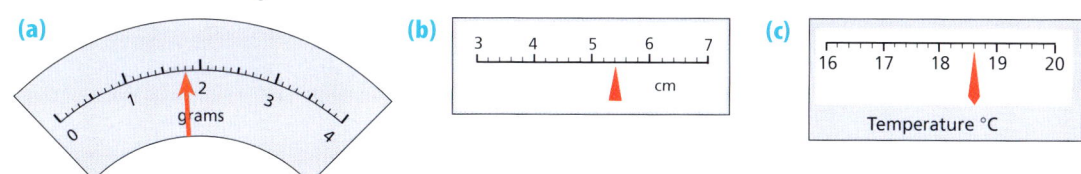

C3 What number does each arrow point to?

(a) 1 2 3

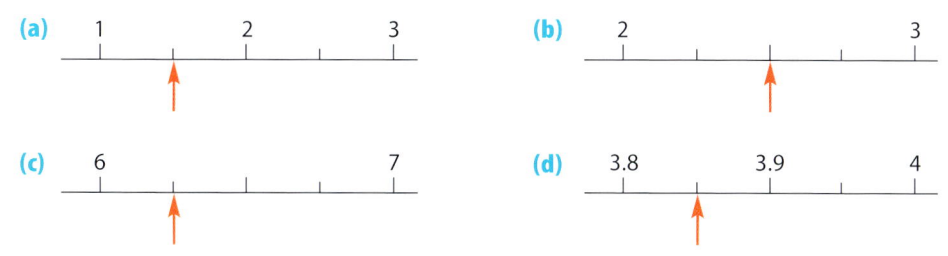

(b) 2 3

(c) 6 7

(d) 3.8 3.9 4

21 Estimating and scales 93

C4 For each of the scales on sheet FT–17 B, draw an arrow to show the reading given.

C5 The graph below shows the average monthly rainfall at Blaenau Ffestiniog in Wales. (J, F, M, … stand for January, February, March, …).

Average monthly rainfall at Blaenau Ffestiniog

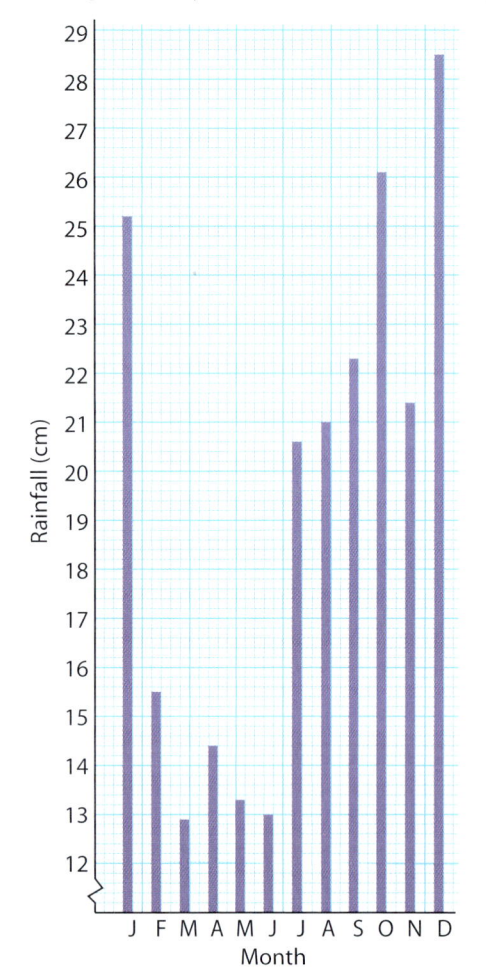

On average, at Blaenau Ffestiniog …

(a) What is the monthly rainfall for
  (i) April
  (ii) February
  (iii) October
  (iv) June
(b) Which are the three wettest months?
(c) How much rain falls in the wettest month?
(d) What is the difference in monthly rainfall between the wettest month and the driest month?

*C6 These scales have no small divisions marked.
Estimate the number that each arrow points to.

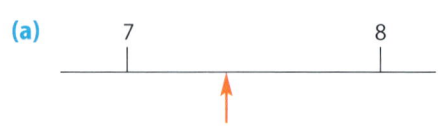

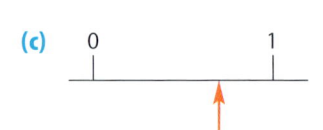

(a) 7 … 8
(b) 1 … 2
(c) 0 … 1
(d) 6.3 … 6.4

94 21 Estimating and scales

Test yourself

T1 Poppy is 1.4 metres tall.
Estimate the height of the door.

T2 The bus in this scale drawing is 4 metres high.

(a) Estimate the height of the lamp-post.
Give your answer in metres.

(b) Two of the instruments in the bus show speed and temperature.

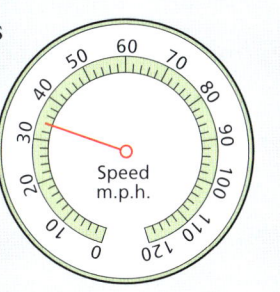

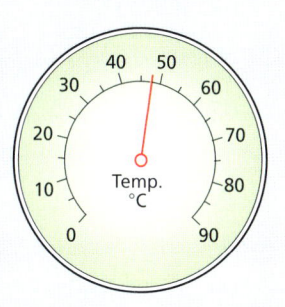

(i) What speed is shown?
(ii) What temperature is shown?

OCR

T3 What is the reading on this scale?

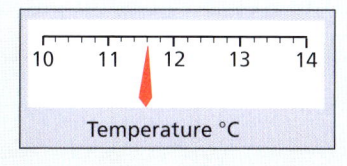

Temperature °C

21 Estimating and scales

22 Multiplying and dividing decimals

This work will help you multiply and divide a decimal by a single digit number.

A Multiplying a decimal by a whole number

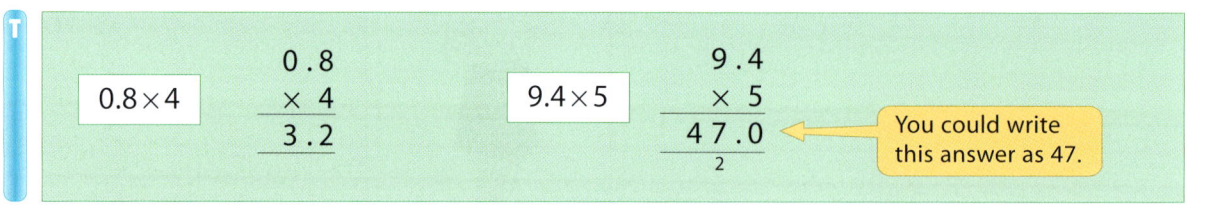

0.8×4

```
  0.8
× 4
———
  3.2
```

9.4×5

```
  9.4
×   5
———
 47.0
   2
```

You could write this answer as 47.

A1 Work these out in your head.
(a) 0.2×9 (b) 0.1×5 (c) 0.6×3 (d) 0.7×2
(e) 1.5×2 (f) 2.5×3 (g) 1.1×4 (h) 2.2×3

A2 Find three matching pairs of multiplications that give the same answer. Which is the odd one out?

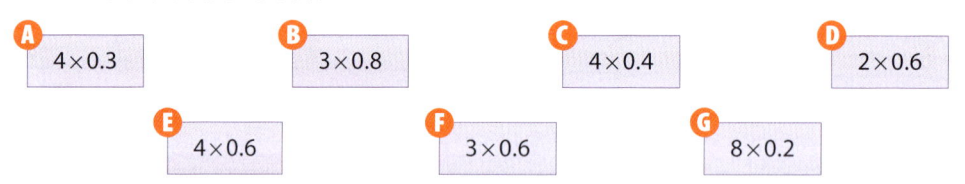

A 4×0.3 **B** 3×0.8 **C** 4×0.4 **D** 2×0.6

E 4×0.6 **F** 3×0.6 **G** 8×0.2

A3 Work these out.
(a) 4.3×5 (b) 8.1×2 (c) 7.4×3 (d) 1.9×7
(e) 5.6×8 (f) 3.9×5 (g) 9.8×4 (h) 7.5×6

A4 Choose numbers from the loop to make these multiplications correct. You can use each number more than once.
(a) ■ × 4 = 0.8 (b) 1.4 × ■ = 4.2
(c) 0.3 × ■ = 1.5 (d) ■ × 0.8 = 4
(e) ■ × 4 = 2.4 (f) 5 × ■ = 1

0.2   0.3   5
0.6   0.5   3

A5 A bag of potatoes weighs 2.5 kg.
What is the weight of 6 of these bags of potatoes?

A6 A book is 2.2 cm thick.
How thick is a pile of 8 of these books?

B Dividing a decimal by a whole number

$14.7 \div 3$

$4\,.9$
$3\overline{)1\,4\,.^{2}7}$

$17 \div 5$

$3\,.4$
$5\overline{)1\,7\,.^{2}0}$

To divide, you sometimes need to put in the decimal point and a zero.

B1 Work these out in your head.
(a) $8.4 \div 2$ (b) $9.3 \div 3$ (c) $12.4 \div 2$ (d) $35.5 \div 5$

B2 Work these out.
(a) $9.2 \div 4$ (b) $7.5 \div 5$ (c) $19.2 \div 6$ (d) $34.8 \div 4$

B3 Choose numbers from the loop to make these divisions correct.
You can use each number more than once.
(a) ■ ÷ 6 = 0.4   (b) 3 ÷ ■ = 0.6
(c) 3 ÷ ■ = 1.5   (d) ■ ÷ 3 = 0.5
(e) ■ ÷ 4 = 0.6   (f) 4.5 ÷ ■ = 0.9

(2.4   1.2   1.5   2   3   5)

B4 Work these out. Write each answer as a decimal.
(a) $28 \div 8$ (b) $27 \div 5$ (c) $38 \div 4$ (d) $63 \div 5$

B5 Find three matching pairs of divisions that give the same answer.
Which is the odd one out?

A $24.6 \div 3$ **B** $28.5 \div 5$ **C** $25.6 \div 4$ **D** $17.1 \div 3$
E $40.8 \div 6$ **F** $32.8 \div 4$ **G** $38.4 \div 6$

B6 A piece of wood 2.5 m long is cut into five equal pieces.
How long is each piece of wood?

Test yourself

T1 Work these out.
(a) 0.5×7 (b) 1.4×3 (c) 3.8×6 (d) 7.5×8

T2 Work these out.
(a) $0.8 \div 4$ (b) $1.6 \div 2$ (c) $9.5 \div 5$ (d) $7.2 \div 6$

T3 A carton contains 1.5 litres of juice.
How much juice is there in 7 of these cartons?

23 Area and perimeter

You will revise finding and estimating the area of a shape on a grid of squares.

This work will help you
- find the area of a rectangle and a right-angled triangle
- find the area of a shape made up of simpler shapes
- find the perimeter of a shape

You need sheet FT–18.

A Shapes on a grid of centimetre squares   level 4

This square has an area of one square centimetre.
We can write this as 1 cm².

A1 Find the area of each of these shapes in square centimetres.

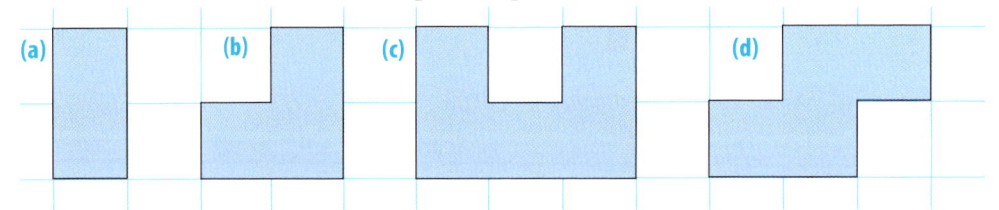

(a) (b) (c) (d)

A2 These shapes have some half squares.
Find the total area of each shape in square centimetres.

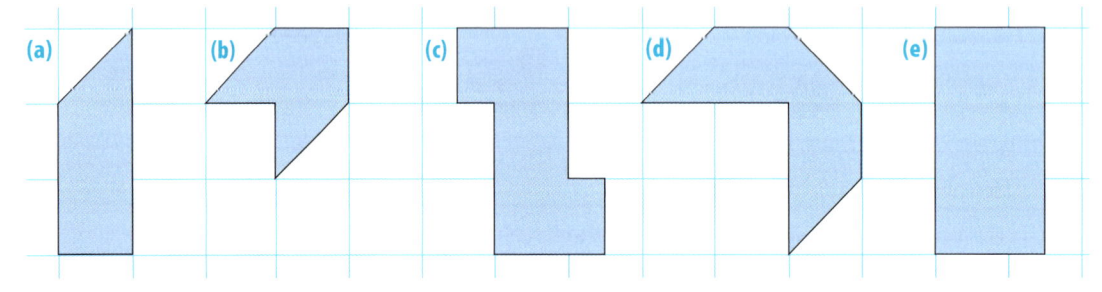

(a) (b) (c) (d) (e)

A3 Write the letters for these shapes in order of their area, smallest first.

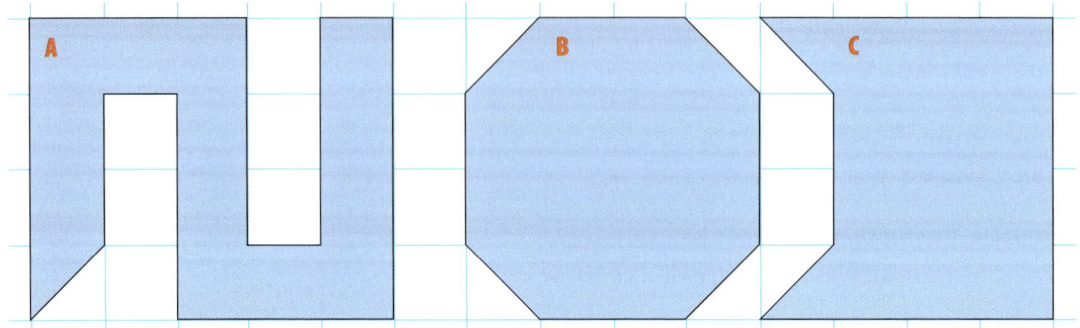

A   B   C

98   23 Area and perimeter

If a curved or irregular shape is drawn on a grid, you can **estimate** its area using a method like this.

If a square has only a small part shaded, do not count it.

If a square is about half shaded, count it as a half square.

If a large part of the square is shaded, count it as one square.

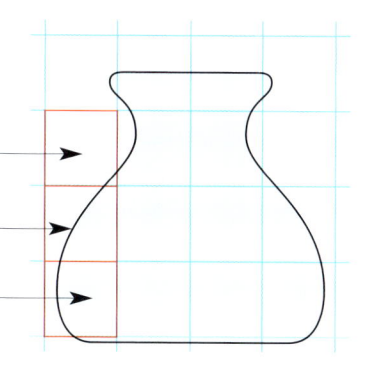

(And remember to count squares that really are whole squares.)

When you have decided which count as one square and which count as a half square, you can set out your working like this.

Estimated total area = 8 full squares + 4 half squares
$= 8 \text{ cm}^2 + 2 \text{ cm}^2$
$= 10 \text{ cm}^2$

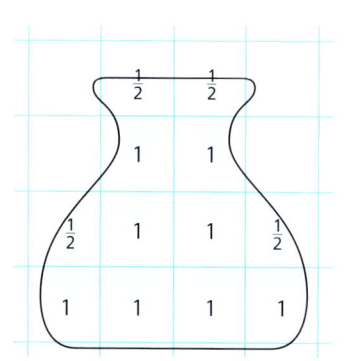

When you estimate an area your answer may be different from someone else's. That does not mean one of you is wrong, but the estimates should be reasonably close.

A4 Find the area of each shape on sheet FT–18.

The **perimeter** of a shape is the distance round the outside of it.

A5 Find the perimeter of each of the shapes in question A1. Remember that the answers must be in centimetres.

A6 Check that this shape has an area of 6 cm² and a perimeter of 12 cm. On centimetre squared paper try to draw

(a) a shape with the same area as here but with a larger perimeter

(b) a shape with the same area as here but a smaller perimeter

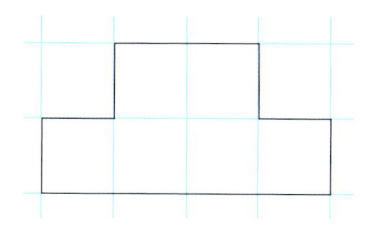

A7 Try to draw

(a) a shape with the same perimeter as here but a larger area

(b) a shape with the same perimeter as here but a smaller area

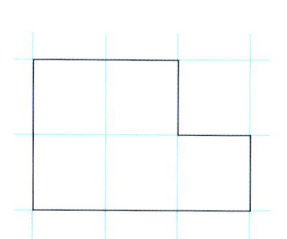

A8 On centimetre squared paper try to draw each of these.

(a) A shape with area 4 cm² and perimeter 8 cm

(b) A shape with area 5 cm² and perimeter 12 cm

(c) A shape with area 8 cm² and perimeter 14 cm

B Area of rectangle and right-angled triangle

B1 Imagine this rectangle split into rows of centimetre squares.

(a) How many rows would there be?

(b) How many squares in each row?

(c) What is the area of the rectangle?

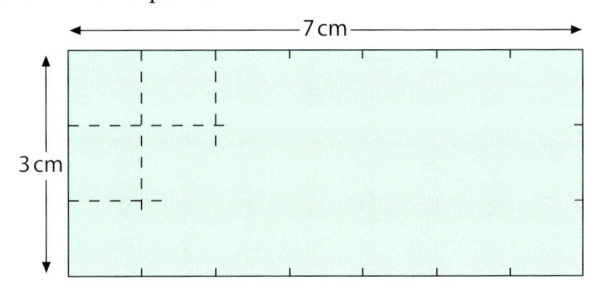

B2 What are the areas of these rectangles in square centimetres?

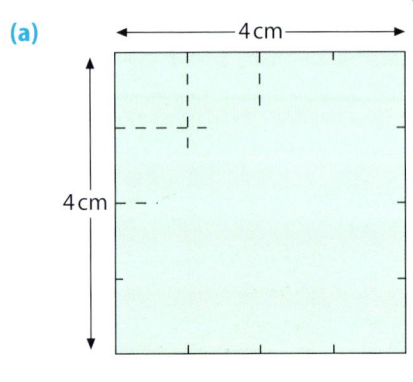

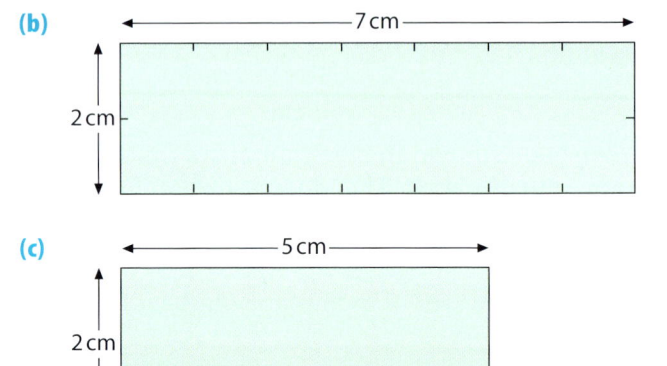

(a)

(b)

(c)

B3 Measure the sides of this rectangle in centimetres.

(a) What numbers do you multiply to work out its area?

(b) What is its area?

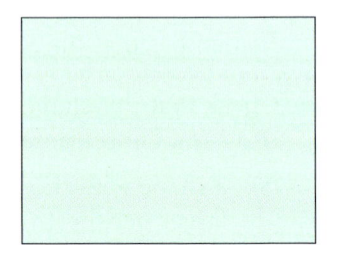

B4 For each of these rectangles, measure the sides, then calculate and write down the rectangle's area.

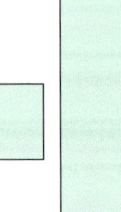

(a)

(b)

(c)

(d)

100   23 Area and perimeter

B5 These are wall tiles. They are not drawn full size.
Calculate the area of each one.

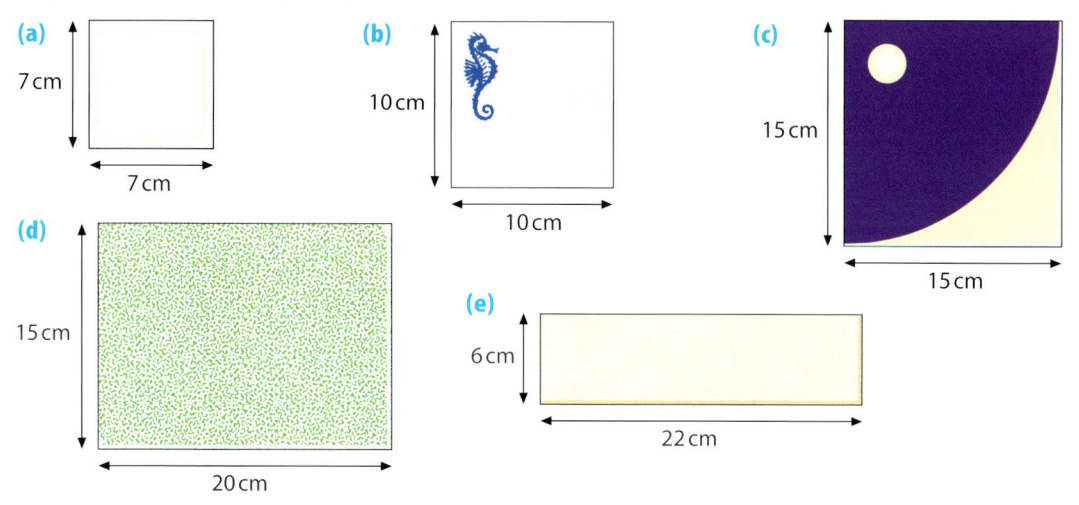

(a) 7 cm × 7 cm

(b) 10 cm × 10 cm

(c) 15 cm × 15 cm

(d) 15 cm × 20 cm

(e) 6 cm × 22 cm

B6 A certain rectangle has an area of 24 cm².
It is 8 cm long. What is its width?

B7 A certain rectangle has an area of 72 cm².
Its sides are whole numbers of centimetres long.
What length could its sides be?
Give as many possibilities as you can.

B8 This rectangle has been divided into two right-angled triangles.

(a) Calculate the area of the whole rectangle.

(b) What is the area of just one of the triangles?

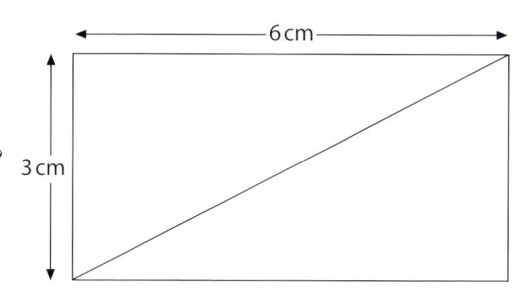

6 cm × 3 cm

B9 In each of these,

(i) calculate the area of the rectangle

(ii) give the area of the blue triangle

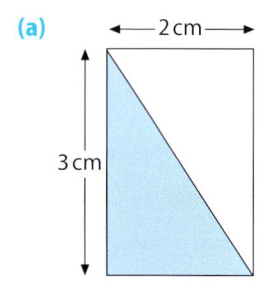

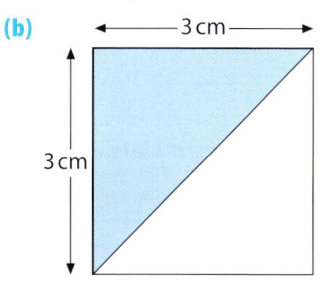

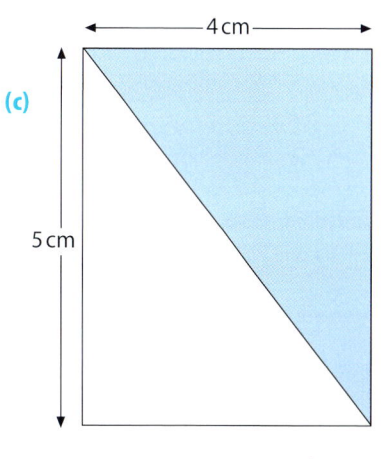

(a) 2 cm × 3 cm

(b) 3 cm × 3 cm

(c) 4 cm × 5 cm

23 Area and perimeter 101

C Area of a shape made from simpler shapes

C1 This L-shape can be split into two rectangles, A and B.

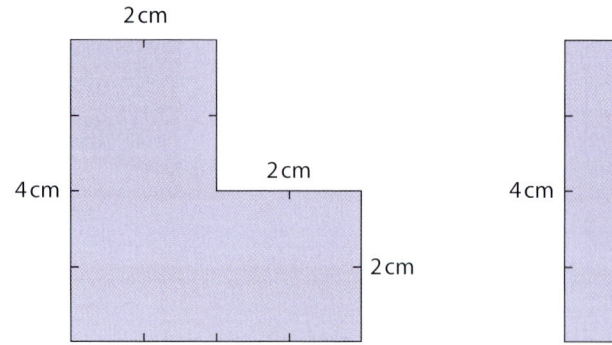

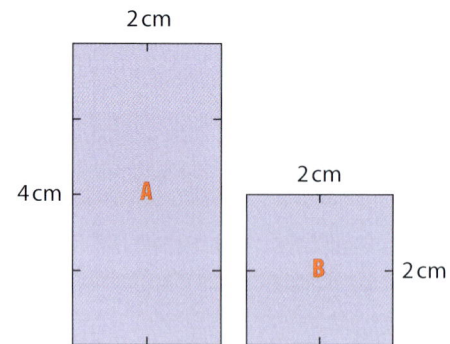

(a) What is the area of rectangle A?
(b) What is the area of rectangle B?
(c) What is the area of the whole L-shape?

C2 This shape has been split into two rectangles.
(a) What is the area of A?
(b) What is the area of B?
(c) Work out the area of the whole shape.

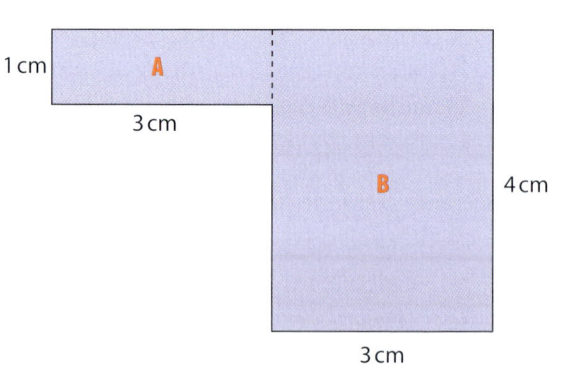

C3 Work out the areas of these L-shapes.
(a)
(b)

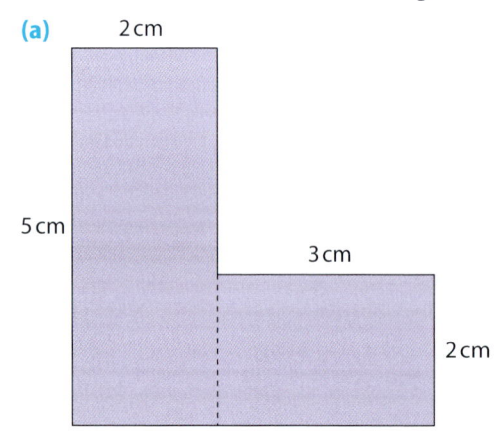

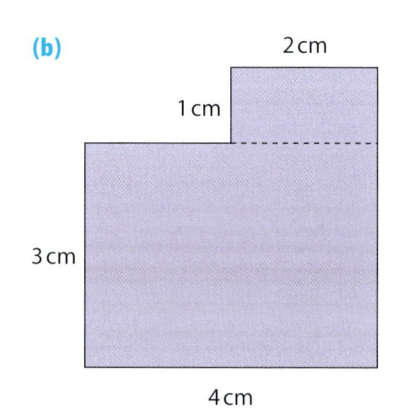

You can usually split up a shape in more than one way.

For example, you can split this L-shape …

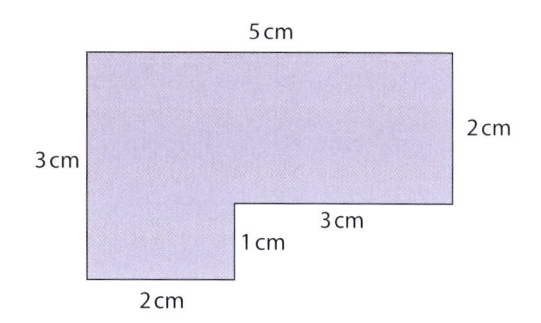

…like this:

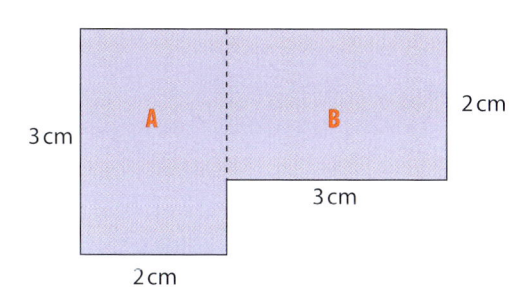

… or like this:

C4 (a) Work out the area of A and then B. What is the area of the L-shape?

   (b) Work out the area of C and then D.
   Check that you get the same area for the whole L-shape.

   (c) Find the perimeter of the L-shape.

C5 (a) Do two sketches of this L-shape.
   Split each sketch in a different way.

   (b) Find the area of the L-shape from one of your sketches.
   Check that the other sketch gives you the same area.

   (c) Find the perimeter of the whole shape.

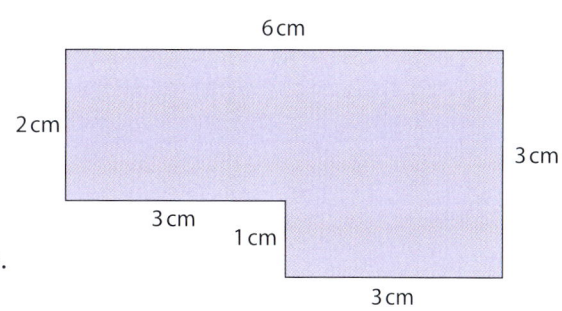

C6 For each of these shapes,

   (i) make a sketch, marking on any missing lengths you may need
   (ii) split the shape into rectangles to find its total area
   (iii) find the perimeter of the whole shape.

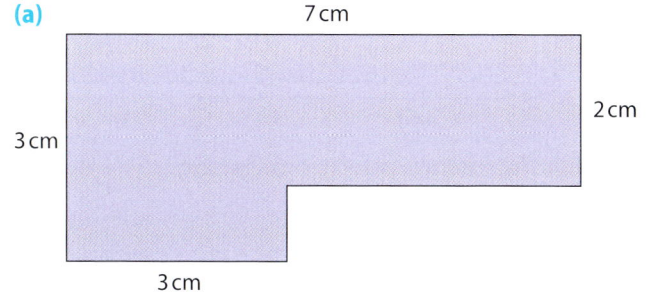

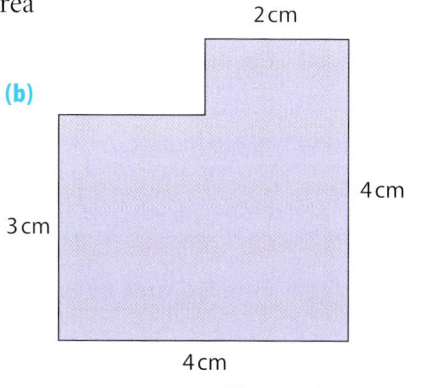

   (a)

   (b)

A square that measures 1 metre by 1 metre has an area of 1 square metre ($1\,m^2$).
We use square metres for the area of a larger surface like the floor of a room.

C7 The diagram below shows the plan of a classroom.

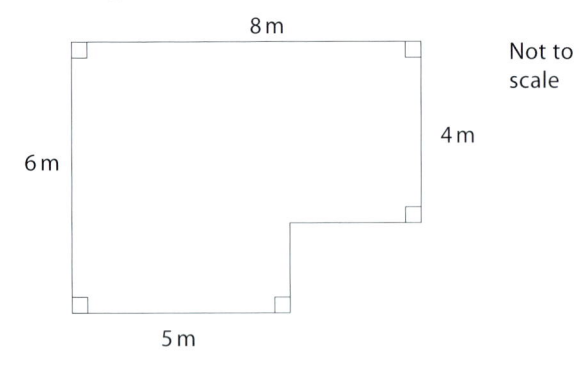

8 m
Not to scale
4 m
6 m
5 m

Work out the perimeter and area of the classroom floor.

OCR

D Using decimals

This is the plan of a flat.
The area of the living room floor is $4.5 \times 3 = 13.5\,m^2$.

4.5 m
3 m

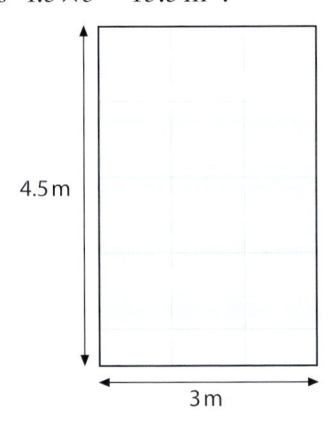

3 m 3.5 m
bedroom kitchen
3.5 m 2 m 3.5 m
WC 1.5 m
4.5 m living room 3 m
hall bathroom
3 m 1.5 m 2 m

D1 Find the area of the floor in
  (a) the bathroom
  (b) the WC
  (c) the bedroom

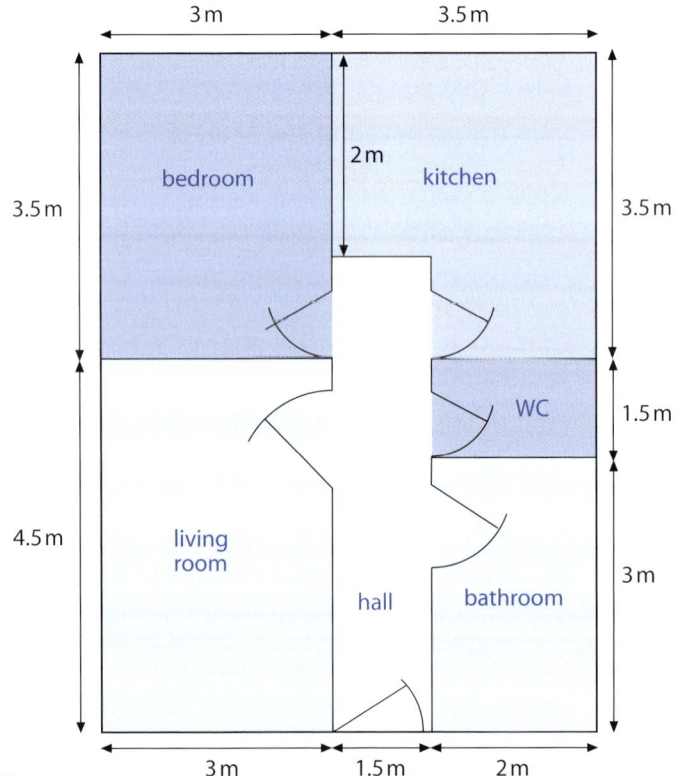

D2 (a) Draw a sketch of the kitchen.
On your sketch draw a line to split the kitchen into two rectangles.
  (b) Find the area of each of these rectangles.
  (c) What is the area of the floor in the kitchen?

The large rectangle measures 2 m by 2.4 m.

Each of these strips is $\frac{1}{10}$ or 0.1 m^2.

There are 4 whole square metres and 8 strips of 0.1 m^2.

So the area of the rectangle is 4.8 m^2.

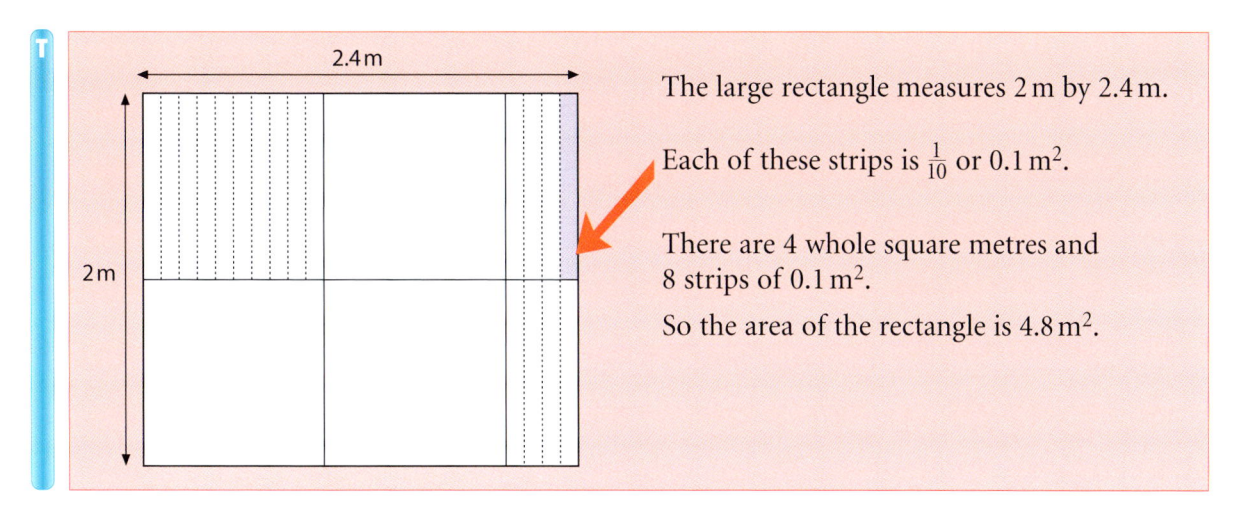

D3 Use your calculator to check the area of the rectangle above.

D4 Here is the plan of another flat.

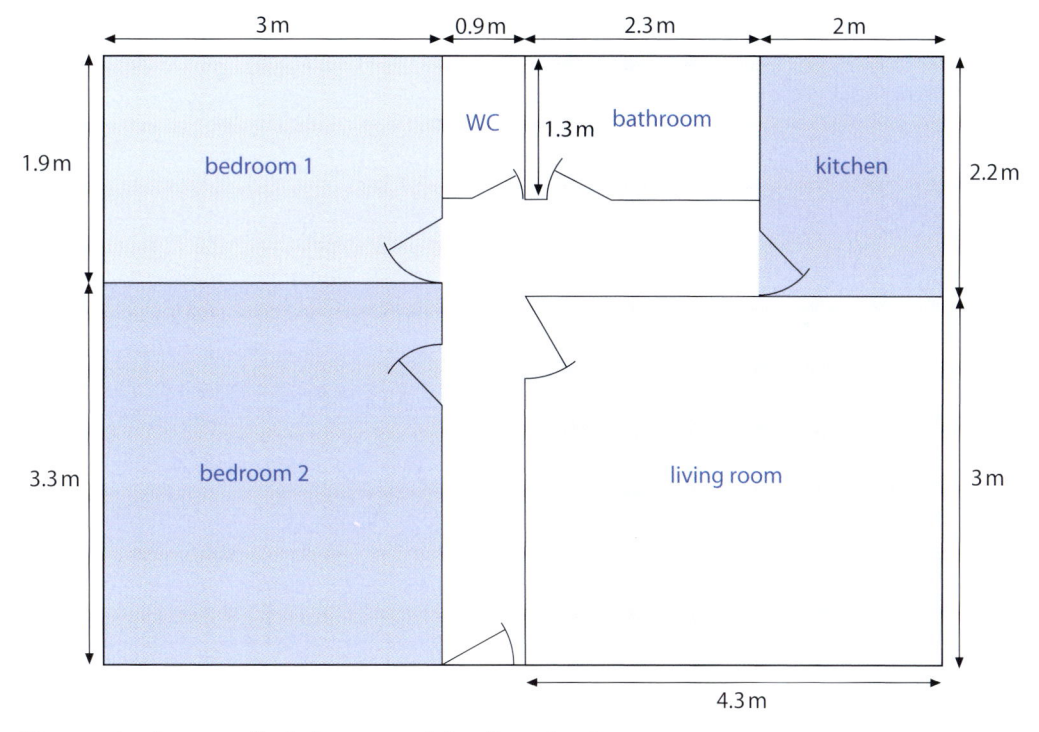

Use a calculator to find the area of the floor in these rooms.

(a) Bedroom 1   (b) Bedroom 2   (c) The living room   (d) The kitchen

D5 Use your calculator to find the area of the WC exactly.

D6 Use your calculator to find the area of the bathroom.

D7 For each of these,
- **(i)** find the area of the rectangle
- **(ii)** use this result to find the area of the blue right-angled triangle.

(a) 3.5 cm, 4 cm

(b) 2.5 cm, 6 cm

*D8 Use a calculator to find the area of each of these right-angled triangles. They are not drawn to scale.

(a) 2.5 cm, 3.6 cm

(b) 5.5 cm, 2.4 cm

Test yourself

T1

These shapes are drawn on a 1 cm grid.
- **(a)** Work out the area of the rectangle.
- **(b) (i)** Work out the area of the T-shape.
  - **(ii)** Work out the perimeter of the T-shape.
- **(c)** Work out an estimate of the area of the irregular shape.

OCR

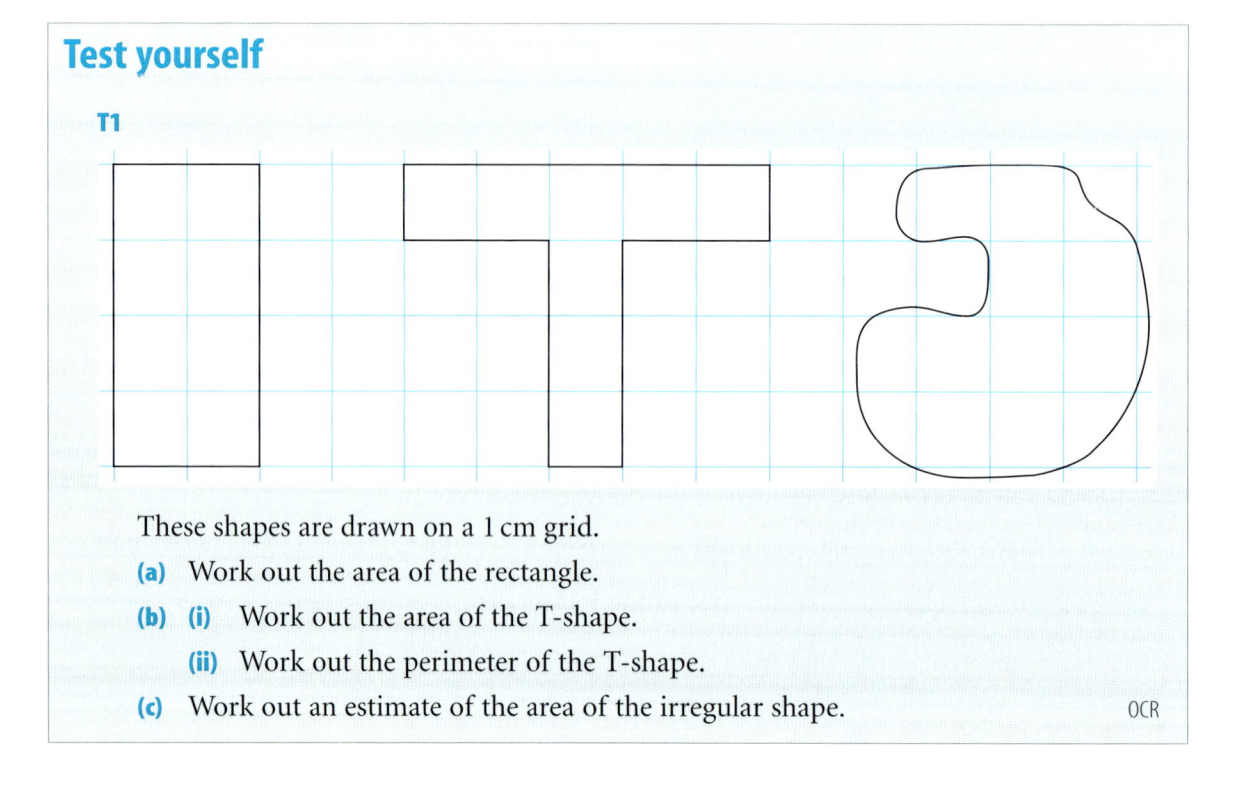

T2 Calculate the area of this shape.
You must show all your working.

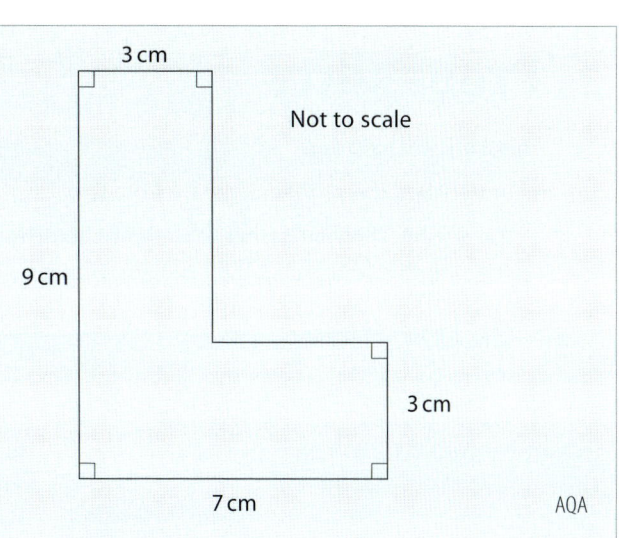

3 cm

Not to scale

9 cm

3 cm

7 cm

AQA

T3 Calculate the perimeter of the shape in question T2.

T4 Calculate the missing lengths in these rectangles.
They are not drawn to scale.

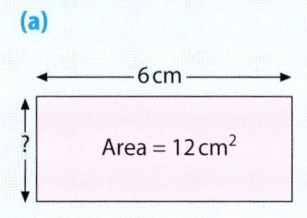

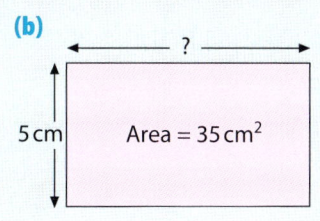

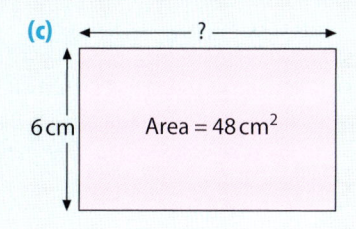

(a) 6 cm, ?, Area = 12 cm²

(b) ?, 5 cm, Area = 35 cm²

(c) ?, 6 cm, Area = 48 cm²

T5 Calculate the area of each of these shapes.

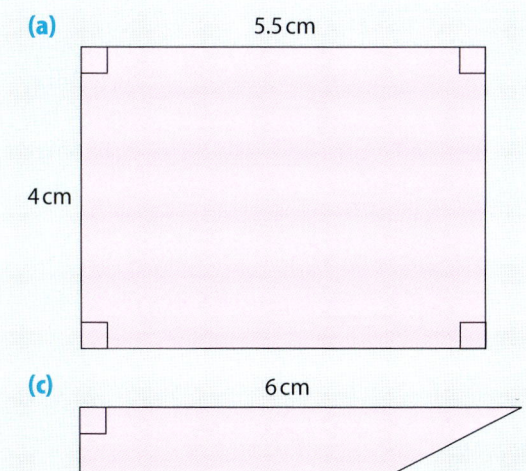

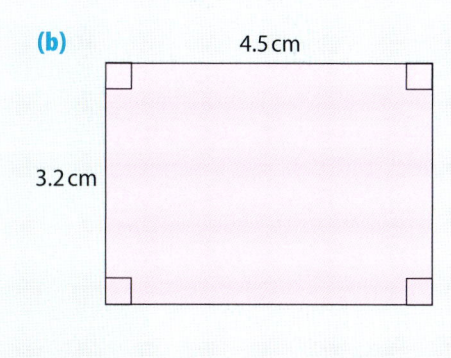

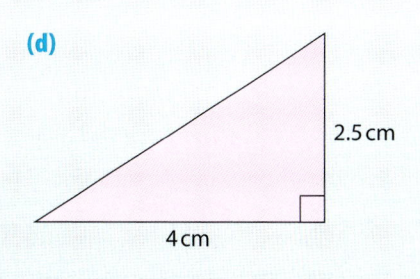

(a) 5.5 cm, 4 cm

(b) 4.5 cm, 3.2 cm

(c) 6 cm, 3 cm

(d) 2.5 cm, 4 cm

24 Probability

This work will help you
- understand the probability scale from 0 to 1
- use equally likely outcomes to find probabilities

A The probability scale

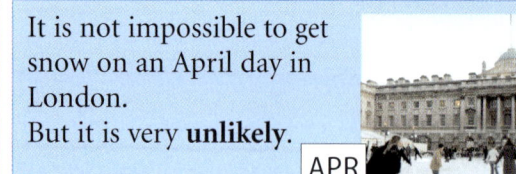

It is not impossible to get snow on an April day in London.
But it is very **unlikely**.

APR 8

Getting a cloudy sky on a January day in London is very **likely**.

JAN 12

Something which is very likely has a high **probability**.
Something which is very unlikely has a low probability.

Probability is measured on a scale that goes from 0 to 1.

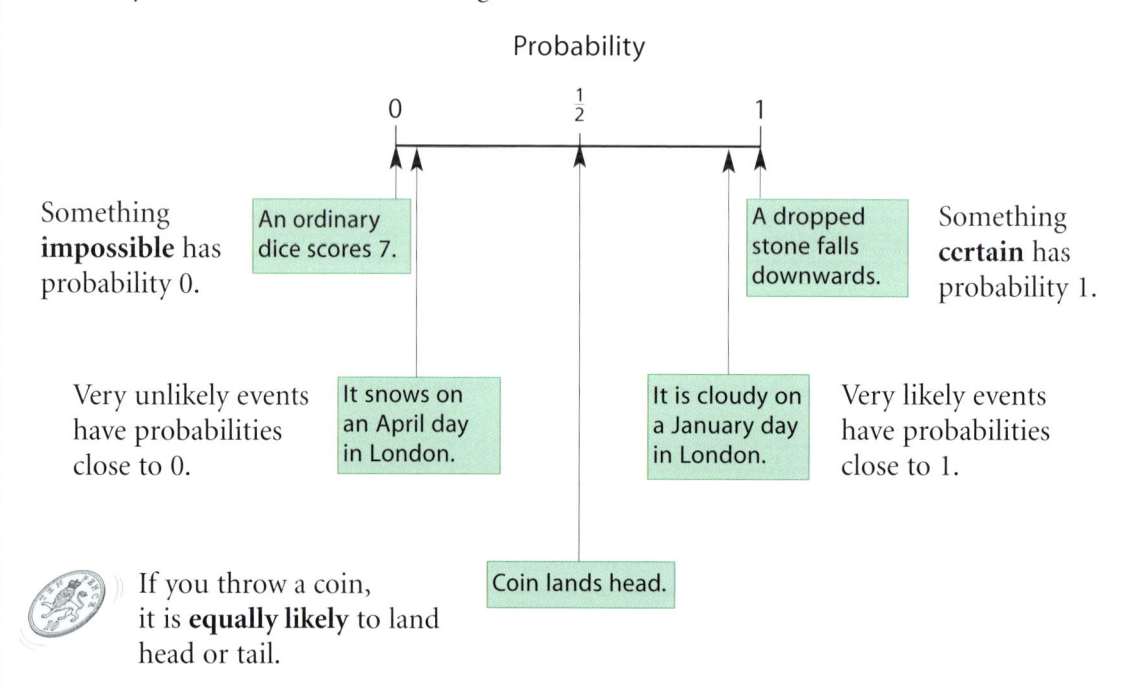

Probability

0     $\frac{1}{2}$ 1

Something **impossible** has probability 0.

An ordinary dice scores 7.

A dropped stone falls downwards.

Something **certain** has probability 1.

Very unlikely events have probabilities close to 0.

It snows on an April day in London.

It is cloudy on a January day in London.

Very likely events have probabilities close to 1.

Coin lands head.

If you throw a coin, it is **equally likely** to land head or tail.

The probability of landing head is $\frac{1}{2}$.

(We also say that there is an **even chance** or a **fifty-fifty chance** of head or tail.)

Picking at random

Donna has some cards.
She shuffles the cards and then picks one without looking.

This is called picking a card **at random**.
Every card is equally likely to be picked.

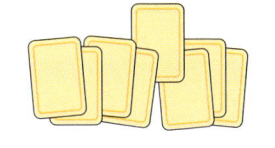

A1 Graham has these cards.

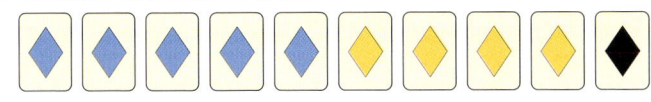

He picks a card at random.

(a) Which colour is he most likely to pick?

(b) Which colour is he least likely to pick?

(c) The probability of each colour is marked on this scale.
What colour does each letter stand for?

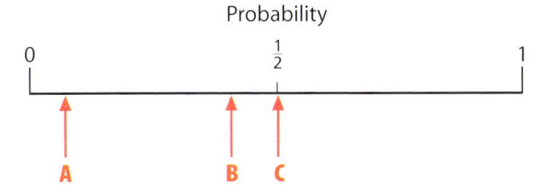

A2 Nadia puts these counters into a bag and shakes them up.

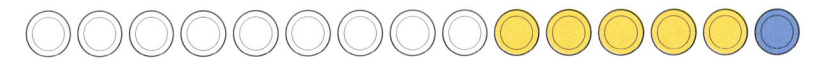

She takes out a counter at random.

(a) Is the probability of taking a white counter less than $\frac{1}{2}$ or more than $\frac{1}{2}$?

(b) Draw a probability scale from 0 to 1 and mark roughly the probability of each colour.

A3 Dave puts these counters into a bag and shakes them up.

He takes out a counter at random.

What is missing from each of these sentences?

(a) Dave is least likely to take a ……… counter.

(b) The chance of taking out a ……… counter is fifty-fifty.

(c) The probability of taking out a white counter is …

A4 Which arrow on the probability scale goes with each statement below?

(a) It is very likely to be sunny tomorrow.

(b) The last train is always late.

(c) Sam is unlikely to win the race.

(d) There is an even chance of a wet or dry day tomorrow.

(e) Earthquakes never happen here.

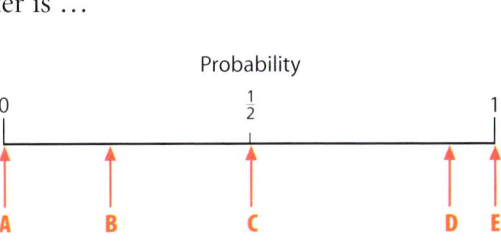

B Equally likely outcomes

Gemma picks a card at random from this set.
Each card is equally likely to be picked.

There are 5 cards. The probability of picking each card is 1 out of 5.
This is written as a fraction: $\frac{1}{5}$.

So the probability of picking red is $\frac{1}{5}$, the probability of picking blue is $\frac{1}{5}$, and so on.

This set contains 7 cards altogether.
Hannah picks a card at random.

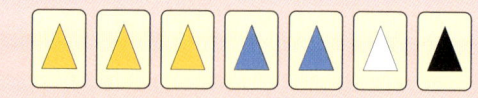

The probability of picking each card is $\frac{1}{7}$.

There are 3 yellow cards out of 7.
So the probability of picking a yellow card is $\frac{3}{7}$.

← 3 yellow
← 7 altogether

Hannah is playing a game with the 7 cards.
She wins if she picks yellow or blue.

There are 5 winning cards 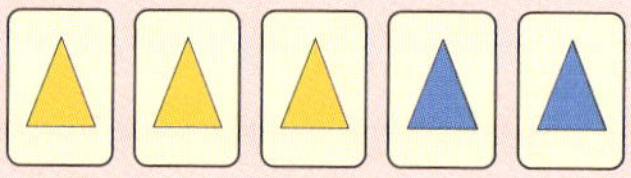 out of 7 altogether.

So the probability of winning is $\frac{5}{7}$.

B1 Roger picks a card at random from this set. 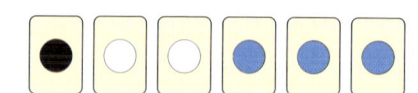
  (a) What is the probability that he picks
   (i) black (ii) white (iii) blue
  (b) Draw a probability scale.
   On it mark the probability of picking each of the three colours.
  (c) Roger plays a game. He wins if he picks either blue or black.
   (i) How many winning cards are there?
   (ii) What is the probability that Roger wins?

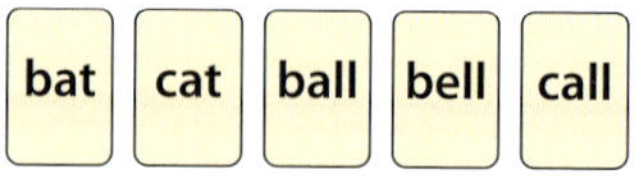

B2 These five cards are placed face down on the table and shuffled about.
Leila picks a card at random.

| bat | cat | ball | bell | call |

What is the probability that she picks
  (a) the word 'bell'
  (b) a word beginning with 'b'
  (c) a word ending in 't'
  (d) a word with both 'a' and 'c' in it
  (e) a word with either 'a' or 'b' in it

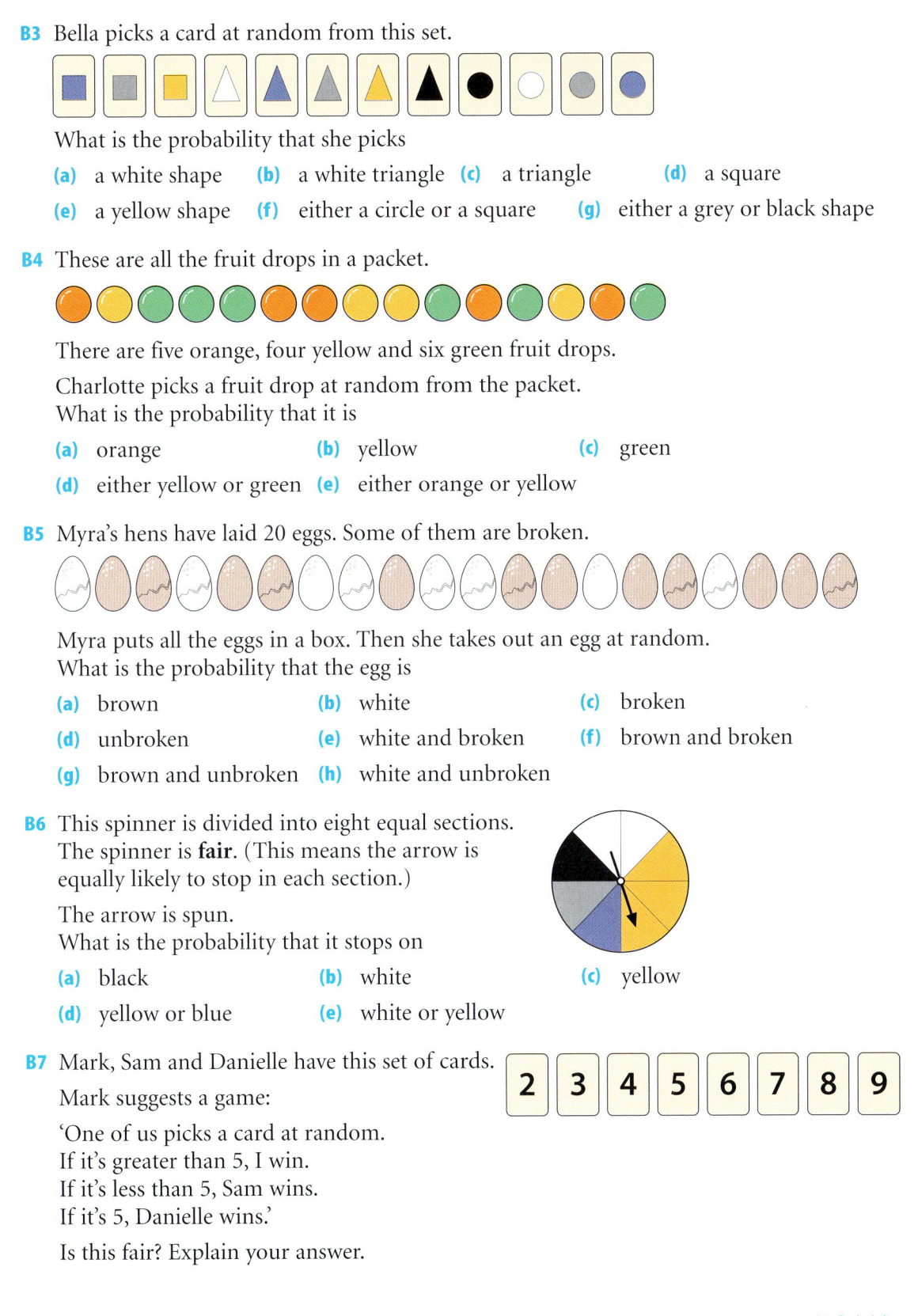

B3 Bella picks a card at random from this set.

What is the probability that she picks

(a) a white shape (b) a white triangle (c) a triangle (d) a square

(e) a yellow shape (f) either a circle or a square (g) either a grey or black shape

B4 These are all the fruit drops in a packet.

There are five orange, four yellow and six green fruit drops.

Charlotte picks a fruit drop at random from the packet.
What is the probability that it is

(a) orange (b) yellow (c) green

(d) either yellow or green (e) either orange or yellow

B5 Myra's hens have laid 20 eggs. Some of them are broken.

Myra puts all the eggs in a box. Then she takes out an egg at random.
What is the probability that the egg is

(a) brown (b) white (c) broken

(d) unbroken (e) white and broken (f) brown and broken

(g) brown and unbroken (h) white and unbroken

B6 This spinner is divided into eight equal sections.
The spinner is **fair**. (This means the arrow is
equally likely to stop in each section.)

The arrow is spun.
What is the probability that it stops on

(a) black (b) white (c) yellow

(d) yellow or blue (e) white or yellow

B7 Mark, Sam and Danielle have this set of cards.

Mark suggests a game:

'One of us picks a card at random.
If it's greater than 5, I win.
If it's less than 5, Sam wins.
If it's 5, Danielle wins.'

Is this fair? Explain your answer.

Test yourself

T1 A fair spinner has three equal sections.
One section is red, one is blue and one is yellow.

The spinner is spun once.
The probabilities of three events have been marked on the probability scale below.

A The spinner lands on blue.

B The spinner lands on green.

C The spinner does **not** land on red.

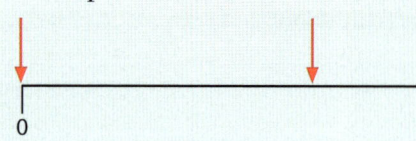

Copy the diagram and label each arrow with the letter to show which event it represents.

AQA

T2 A fair dice is thrown.

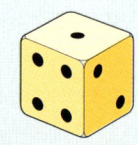

The probabilities of five events have been marked on a probability scale.

V A 5 is thrown.

W An odd number is thrown.

X A 7 is thrown.

Y A number less than 7 is thrown.

Z A 2 is **not** thrown.

Which event does each arrow show?

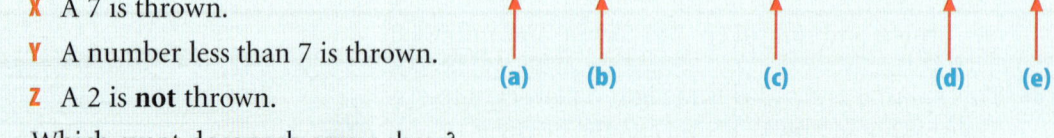

T3 This fair spinner has eight equal sections.
One section is blue, two are green and the others are red.

The spinner is spun once.
What is the probability that it stops on

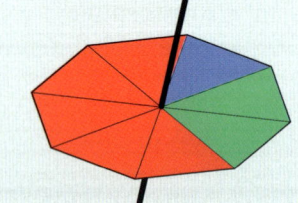

(a) blue

(b) green

(c) blue or green

(d) red or blue

(e) yellow

Review 4

1. Look at this number pattern.

    2   7   12   17   22   27   …

    (a) (i) What is the next number in the pattern?

    (ii) Explain how you worked out your answer.

    (b) Is 100 a number in this number pattern?
    Explain how you decided.

2. Five identical cars are parked in a line, bumper to bumper.
   How long is the line of cars if each one is 3.4 metres long?

3. What temperature is shown on each of these scales?

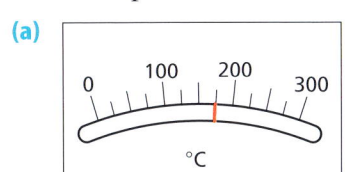

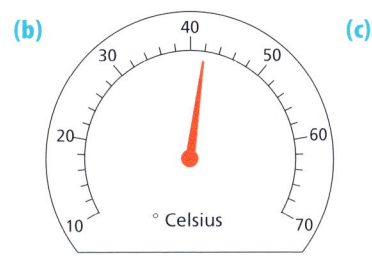

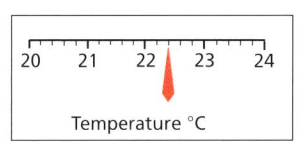

    (a)   (b)   (c)

4. 6 kg of cheese is divided into four equal pieces. How much does each piece weigh?

5. Find the area of each shape (not drawn to scale).

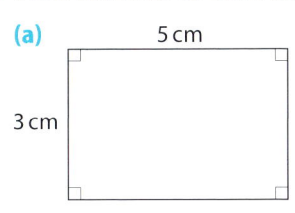

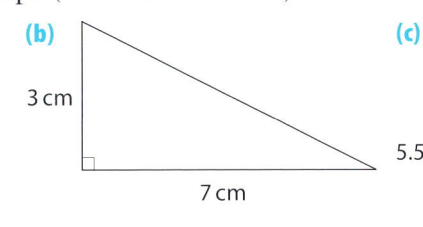

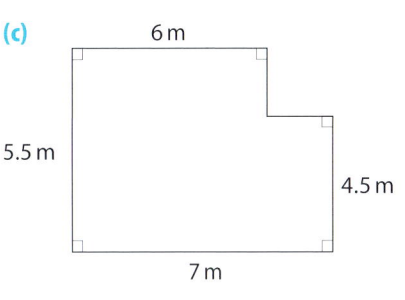

    (a) 5 cm × 3 cm

    (b) triangle, 3 cm and 7 cm

    (c) 6 m, 5.5 m, 4.5 m, 7 m

6. What is the probability of rolling a 5 when an ordinary dice is rolled once?

7. Work these out.

    (a) 1.9×3 (b) 6×0.7 (c) $14.8 \div 4$ (d) $12 \div 5$

8. What is the next number in this sequence? 0.3, 0.6, 0.9, 1.2, 1.5, …

9. This fair spinner is spun once.
   What is the probability that it stops on

    (a) blue   (b) orange

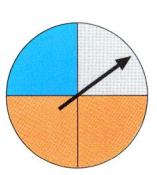

10. What are the next two numbers in this sequence? 1, 3, 7, 13, 21, …

25 Enlargement

This work will help you
- enlarge a shape by a given scale factor
- work out what scale factor has been used for an enlargement

A Enlargement on squared paper

To make a **two times enlargement** of a shape you double the length of each side.

A1 Make a two times enlargement of this shape on centimetre squared paper. One line has been done for you.

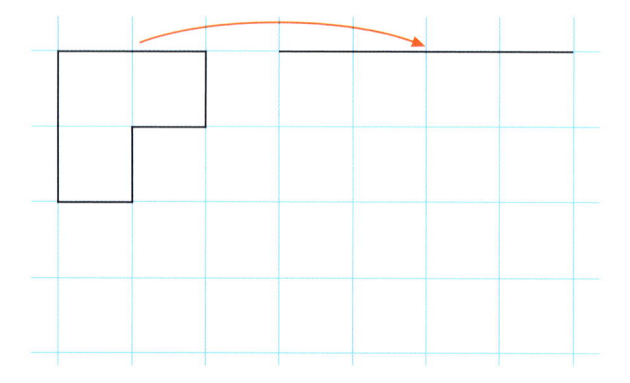

A2 On centimetre squared paper, make a two times enlargement of each of these shapes.

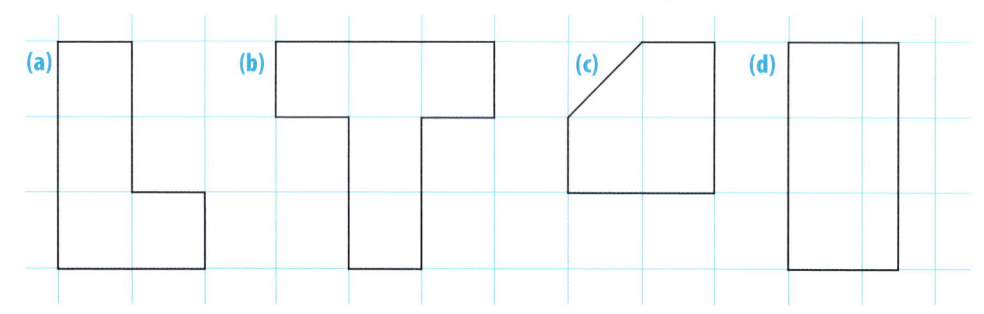

(a) (b) (c) (d)

A3 (a) What do you do to each side to make a **three times enlargement** of a shape?

(b) On centimetre squared paper, make a three times enlargement of each of these shapes.

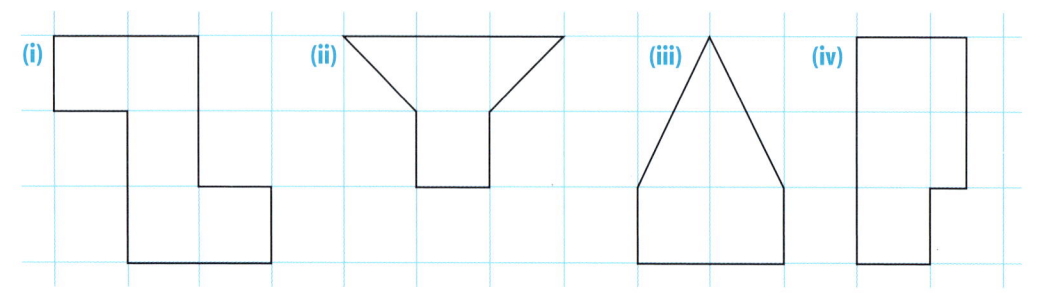

(i) (ii) (iii) (iv)

B Scale factor

When you draw a two times enlargement of a shape, you are enlarging the shape using a **scale factor** of 2.

B1 Which of these are correct enlargements of shape A?
Give the scale factor that has been used for each correct enlargement.

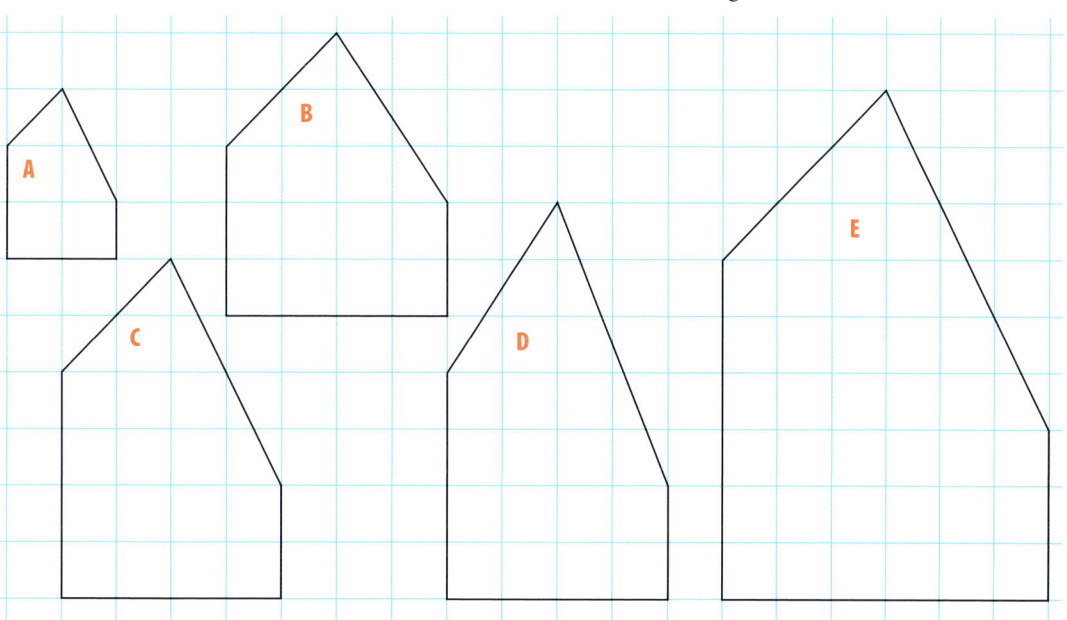

B2 (a) Triangle Q is an enlargement of triangle P.
Measure the base of P and the base of Q, then work out what scale factor has been used for the enlargement.

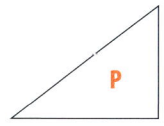

(b) Measure the height of P and the height of Q, then check that the same scale factor has been used.

(c) Triangle R is also an enlargement of triangle P.
Measure the base or height, then work out the scale factor of this enlargement.

25 Enlargement 115

The **perimeter** of a shape is the distance all round it.

B3 Shape X is drawn on centimetre squared paper.

(a) Work out the perimeter of shape X.

(b) On centimetre squared paper, draw an enlargement of shape X with scale factor 2. What is the perimeter of this enlargement?

(c) Now draw an enlargement of X with scale factor 3. What is the perimeter of this enlargement?

(d) What happens to the perimeter of a shape when it is enlarged?

B4 This is an enlargement of a smaller rectangle.
A scale factor of 2 has been used.

(a) What was the width of the smaller rectangle?

(b) What was the height of the smaller rectangle?

(c) What was the perimeter of the smaller rectangle?

B5 A shape is enlarged by scale factor 5.
The width of the enlargement is 15 cm.
What was the width of the original shape?

B6 A shape has perimeter 8 cm.
An enlargement of the shape has a perimeter of 32 cm.
What is the scale factor of this enlargement?

Test yourself

T1 Copy this shape on to centimetre squared paper.
Draw an enlargement of the shape.
Use a scale factor of 2.

OCR

T2 Shapes P and Q are each enlargements of shape X.
Give the scale factor of each enlargement.

26 Negative numbers

This work will help you
- put positive and negative temperatures in order
- solve problems involving temperature changes
- use positive and negative coordinates

A Putting temperatures in order

The table shows the temperatures in some European cities one day in March.

Temperatures are usually measured in degrees Celsius (°C).

The temperature in Moscow is ⁻3 °C.
This means that it is 3 degrees below 0 °C.

City	Temperature
Berlin	7 °C
Copenhagen	2 °C
Helsinki	⁻5 °C
Moscow	⁻3 °C
Oslo	⁻4 °C
Prague	0 °C
Riga	3 °C

- How many cities have temperatures below 0 °C?
- Which city is warmest?
- Which city is coldest?
- Put the temperatures in order, starting with the coldest.

A1 (a) Write down the temperature, in °C, shown on each of these thermometers.

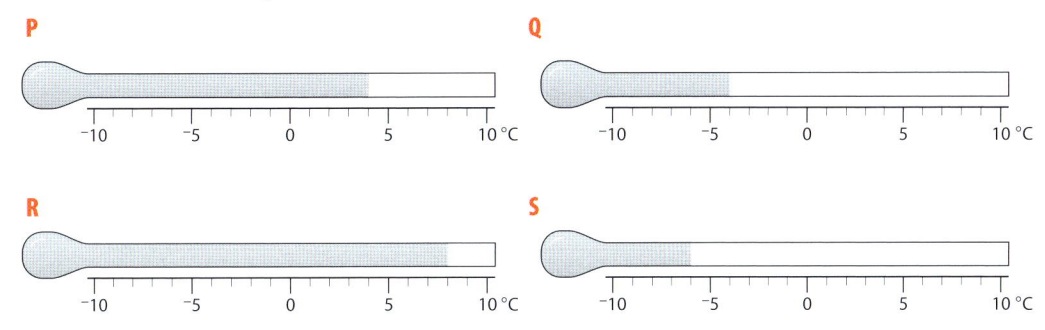

P

Q

R

S

(b) Which thermometer shows the coldest temperature?
(c) Which thermometer shows the warmest temperature?
(d) Write the temperatures in order, starting with the coldest.

A2 Max's car displays the outside temperature.
Max records the outside temperature when he drives to work each morning for a week.

	Mon	Tue	Wed	Thur	Fri
Temperature	⁻2°C	⁻3°C	1°C	3°C	⁻5°C

(a) Which day was warmest?

(b) Which day was coldest?

(c) Which day was closest to 0°C?

(d) Write the temperatures in order, starting with the coldest.

A3 Write these lists of temperatures in order, starting with the lowest.

(a) 0°C, 2°C, ⁻2°C, 3°C, ⁻4°C   (b) 5°C, ⁻10°C, 3°C, ⁻2°C, ⁻6°C

A4 Jasmine measures the temperature every day when she gets up.
One weekend in January the temperature was ⁻5°C on Saturday and ⁻2°C on Sunday.
Which day was colder?

A5 The table shows the temperatures in Aviemore in Scotland for five days in March.

	Mon	Tue	Wed	Thur	Fri
Maximum day	6°C	5°C	7°C	3°C	5°C
Minimum night	1°C	⁻1°C	⁻1°C	⁻2°C	⁻4°C

(a) What was the maximum day temperature on Monday?

(b) What was the minimum night temperature on Thursday?

(c) Which day had the highest temperature?

(d) Which two days had the same maximum day temperature?

(e) Which two days had the same minimum night temperature?

(f) Which day was coldest in the night?

A6 The map shows the temperatures in some cities in December.

(a) Which city was coldest?

(b) Which city was warmest?

(c) Which city was closest to 0°C?

(d) Write the temperatures in order, starting with the lowest.

Glasgow ⁻2°C
Edinburgh ⁻4°C
Manchester ⁻3°C
Birmingham 3°C
Cardiff 4°C
London 5°C

B Temperature changes

The temperature at midnight was ⁻4 °C.
The temperature at midday was 3 °C.
The temperature went up between midnight and midday.

The number line shows that the temperature went up by 7 degrees.

Use the number line to answer these.

- At 7 a.m. it was ⁻2 °C.
  At 1 p.m. it was 2 °C.
  How many degrees warmer was it at 1 p.m.?

- When I woke up on Tuesday it was ⁻4 °C.
  When I woke up on Wednesday it was ⁻1 °C.
  How many degrees colder was it on Tuesday?

- At noon in London it was 5 °C.
  At noon in Paris it was ⁻1 °C.
  How many degrees warmer was it in London?

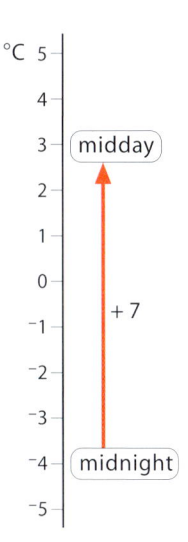

B1 **(a)** Write down the temperature shown on thermometer A.

(b) Write down the temperature shown on thermometer B.

(c) Copy and complete the sentence below.

Temperature … is warmer by … degrees.

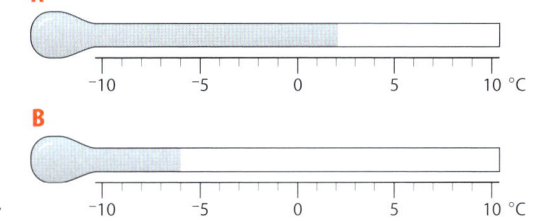

B2 The temperature in Tara's garage was 1 °C.
The temperature outside was ⁻3 °C.

How many degrees colder was it outside than inside the garage?

B3 At 7 a.m. the temperature was ⁻4 °C.
By midday the temperature had gone up by 6 degrees.

What was the temperature at midday?

B4 On Tuesday the temperature at noon was 3 °C.
On Wednesday the temperature at noon was 4 degrees lower.

What was the temperature at noon on Wednesday?

B5 The table shows the temperature recorded at the Cairngorm weather station one day in March 2006.

Time	00:18	04:18	08:18	12:18	16:18	20:18
Temperature	⁻6 °C	⁻5 °C	⁻3 °C	⁻4 °C	⁻4 °C	⁻5 °C

(a) At what time was the temperature highest?

(b) At what time was the temperature lowest?

(c) By how many degrees did the temperature fall between 08:18 and 20:18?

(d) How many degrees warmer was it at 16:18 than at 04:18?

B6 This table shows the temperature on the surface of each of five planets.

Planet	Temperature
Venus	480 °C
Mars	⁻60 °C
Jupiter	⁻150 °C
Saturn	⁻180 °C
Uranus	⁻210 °C

(a) Work out the difference in temperature between Mars and Jupiter.

(b) Work out the difference in temperature between Venus and Mars.

(c) Which planet has a temperature 30 degrees higher than the temperature on Saturn?

The temperature on Pluto is 20 degrees lower than the temperature on Uranus.

(d) Work out the temperature on Pluto.

Edexcel

C Negative coordinates

- What word is spelt by the coordinates (4, 2) (3, 0) (−3, 3) (−3, −2)?
- What are the coordinates for the word SNAKE?
- Write your name using the coordinate code.

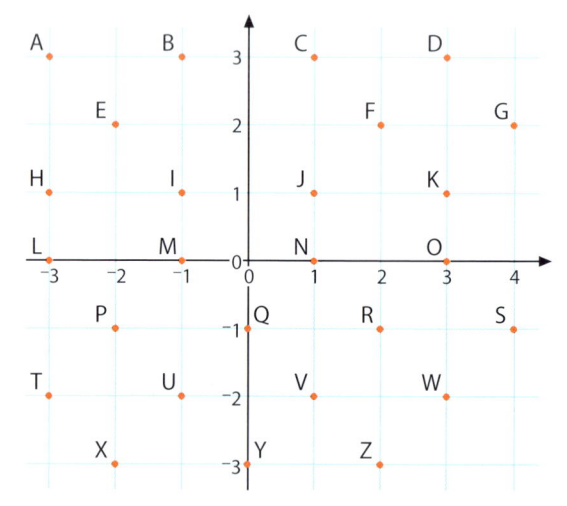

C1 Use the grid above to write down the coordinate codes for each of these.

(a) HERON   (b) SPARROW

(c) PUFFIN   (d) MALLARD

C2 Use the grid to write down the letters given by each set of coordinates below.
Rearrange the letters to give the name of an animal.

(a) (⁻3, 3) (⁻2, 2) (2, ⁻1) (⁻1, 3)

(b) (⁻1, 1) (1, 0) (⁻3, 0) (3, 0)

(c) (2, ⁻1) (⁻2, 2) (⁻1, 1) (⁻3, ⁻2) (4, 2)

(d) (4, 2) (⁻3, 3) (⁻1, 3) (2, ⁻1) (3, 3) (⁻2, 2)

(e) (2, 2) (3, 0) (⁻3, 0) (3, ⁻2)

(f) (2, ⁻1) (⁻3, 3) (⁻3, 3) (1, 1) (⁻1, ⁻2) (4, 2)

Test yourself

T1 Sally wrote down the temperature at different times on 1st January 2003.

Time	Temperature
midnight	⁻6 °C
4 a.m.	⁻10 °C
8 a.m.	⁻4 °C
noon	7 °C
3 p.m.	6 °C
7 p.m.	⁻2 °C

(a) Write down
  (i) the highest temperature
  (ii) the lowest temperature

(b) Work out the difference in temperature between
  (i) 4 a.m. and 8 a.m.
  (ii) 3 p.m. and 7 p.m.

At 11 p.m. that day the temperature had fallen by 5 degrees from its value at 7 p.m.

(c) Work out the temperature at 11 p.m.

Edexcel

T2 (a) The temperatures in a number of cities in Europe on one day in December were

11 °C   5 °C   ⁻2 °C   ⁻7 °C   8 °C   14 °C   ⁻3 °C

(i) Write down the coldest of these temperatures.

(ii) Write down the warmest of these temperatures.

(b) The temperature recorded in Helsinki, Finland, was ⁻3 °C at 2 p.m. At midnight this temperature had fallen by 8 degrees.
What was the temperature at midnight?

AQA

T3 (a) Which letter is at the point
  (i) (2, 1)   (ii) (⁻2, ⁻1)

(b) Write down the coordinates of the letter
  (i) G   (ii) E

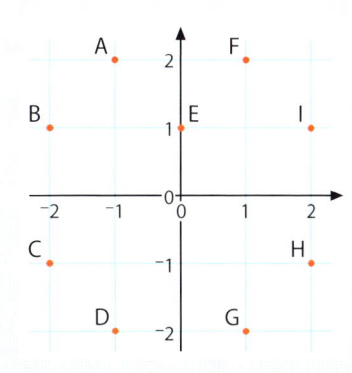

27 Mean

This work will help you
- find the mean of a set of data
- use the mean and range to compare two data sets

A Finding the mean of a data set

Here are the weights in kilograms of the players in a five-a-side football team.

    49   52   39   45   50

To find the **mean** weight, first add up all the weights.

    49 + 52 + 39 + 45 + 50 = 235

Then divide by the number of weights (5).

    235 ÷ 5 = 47

So the **mean** weight is 47 kg.

The mean is an 'average' weight.
Some players weigh more than 47 kg and some less.

The mean does not have to be a whole number.
Here are the weights of another team.

    47   54   61   56   49

The total weight of this team is 47 + 54 + 61 + 56 + 49 = 267.
The mean weight is 267 ÷ 5 = 53.4 kg.

A1 Here are the weights in kilograms of the seven players in a netball team.

    50   64   55   53   48   60   62

(a) Find the total weight of the team.
(b) Find the mean weight.

A2 A sow gives birth to eight piglets.
Four months later the weights, in kg, of the piglets are.

    12   14   10   20   19   14   9   18

(a) Find the total weight of the piglets.
(b) Find the mean weight.

A3 Janis has six snakes. Their lengths, in cm, are 78, 47, 55, 88, 53, 69.
Find the mean length of these snakes.

A4 Gary visits five supermarkets to find the price of the same can of soup.
The five prices are 68p, 55p, 89p, 75p, 90p.

Find the mean of these prices.

A5 Mandy records the temperature at noon each day for a week.
Here are her results.

Day	Mon	Tue	Wed	Thu	Fri	Sat	Sun
Temperature (°C)	14	15	12	10	14	12	14

Find the mean temperature at noon for the week.

A6 Find the mean of each of the following data sets.

(a) 45 23 34 27 47 46
(b) 89 66 90 67 92 68 66 79 67
(c) 3.7 3.8 2.9 2.4 5.1 4.8 4.3 4.4
(d) 28 25 27 35 29 19 26 28 24 30
(e) 237 312 268 305

A7 Each week Marcel has a maths test, marked out of 20.
His marks for the past nine weeks are   14 16 13 10 15 17 16 15 19

(a) Find Marcel's mean mark for the nine tests.

(b) In the next test, Marcel gets 18 marks.
Find the mean mark for all ten tests.

A8 Six friends each do a 'health walk' after work every day for a week.
This table shows the distance, in kilometres, that each person walked on each day.

Name	Mon	Tue	Wed	Thu	Fri
Jess	3.3	2.1	3.0	3.5	3.8
Mary	2.9	2.6	2.1	3.7	3.4
Sean	3.7	3.9	3.4	3.5	2.9
Suki	2.0	2.4	2.7	2.7	2.4
Tom	1.9	2.2	2.5	3.1	3.6
Will	2.3	2.7	2.4	3.3	4.0

Find

(a) Mary's mean distance
(b) Tom's mean distance
(c) Suki's mean distance
(d) the mean distance walked by the six friends on Thursday
(e) the mean distance walked on Friday

B Comparing two sets of data

The mean is often used as an 'average' to compare two sets of data.

For example, here are the weights in grams of the tomatoes picked from two plants, A and B.

Plant A   47   54   59   48   63

Plant B   51   57   52   49   50   55   57

The mean weight for plant A is $\frac{47 + 54 + 59 + 48 + 63}{5} = \frac{271}{5} = 54.2\,\text{g}$.

The mean weight for plant B is $\frac{51 + 57 + 52 + 49 + 50 + 55 + 57}{7} = \frac{371}{7} = 53\,\text{g}$.

So the tomatoes from plant A are heavier, on average.

B1 Two groups of children, boys and girls, collect money for charity.

There are six boys. The amounts they collect are £12 £20 £18 £22 £14 £16.

There are eight girls. The amounts they collect are £10 £13 £19 £25 £24 £17 £20 £14.

(a) Find the mean amount collected by the boys.

(b) Find the mean amount collected by the girls.

(c) Which group collected more, on average?

Reminder: The **range** of a data set is largest value – smallest value.
It is used to compare the spread of two data sets.

B2 (a) Find the range of the boys' amounts in question B1.

(b) Find the range of the girls' amounts.

(c) For which group were the amounts more widely spread out?

B3 This season, Rovers have played eight games. The numbers of goals scored were

5   1   4   2   0   3   3   2

United have played ten games. The numbers of goals scored were

3   0   2   4   4   1   1   3   4   0

(a) Find the mean number of goals in a game for each team.

(b) Which team scored more goals, on average?

B4 A band has a practice every week.
Last term the numbers turning up at the practices were 15, 17, 16, 18, 18, 12, 13, 17.
This term the numbers were 17, 16, 11, 18, 16, 13, 17, 14, 10, 18.

(a) Find the mean number turning up last term.

(b) Find the mean number turning up this term.

(c) In which term were the practices better attended?

B5 There are 11 girls and 9 boys in Gina's class. Here are their heights, in cm.

Girls   141   137   149   136   144   146   139   140   153   150   149
Boys   147   154   148   160   153   139   146   154   158

(a) Find (i) the mean of the girls' heights (ii) the range of the girls' heights
(b) Find (i) the mean of the boys' heights (ii) the range of the boys' heights
(c) Which of the two groups, girls or boys, is taller on average?
(d) Which of the two groups is more spread out in height?

B6 The salaries of the people who work for Company A are as follows.
(£17k means £17 000.)

£17k   £18k   £20k   £23k   £25k   £29k   £34k   £42k

The salaries of the people who work for Company B are

£13k   £15k   £16k   £20k   £26k   £27k   £29k   £36k   £40k   £41k

(a) In which company are salaries higher, on average?
Show how you get your answer.
(b) Which company has the wider spread of salaries?
Show how you get your answer.

B7 Karl counted the eggs in 12 nests of one type of bird. The numbers were

3   4   2   5   4   5   3   2   3   4   2   2

(a) Find the mean number of eggs in a nest.
(b) In 10 nests of a different type of bird, Karl counted 35 eggs altogether.
Find the mean number of eggs in a nest for this type of bird.

Test yourself

T1 Find the mean of each of these sets of data.
(a) 34   38   33   39   54   61   43   42
(b) 26   17   50   34   28   35   50   34   29   33

T2 Gemma has two pepper plants.
The weights, in grams, of the peppers from the two plants are

First plant   56   66   43   67   82   60   32
Second plant   46   35   49   58   88   59   57   52

(a) Find the mean weight of the peppers from
(i) the first plant   (ii) the second plant
(b) Find the range of the weights of the peppers from
(i) the first plant   (ii) the second plant
(c) Which plant produced heavier peppers, on average?
(d) For which plant were the weights more spread out?

28 Starting equations

This work will help you
- find the missing number in an arrow diagram
- solve one-step number puzzles
- solve one-step equations

You need sheets FT–19 and FT–20, and some coloured counters.

A Arrow diagrams
level 3

- Find the missing number at the end of each arrow diagram.

  2 →(+5)→ ?
  10 →(−3)→ ?
  3 →(×4)→ ?
  6 →(÷2)→ ?

- Now find the missing number at the **beginning** of each diagram.

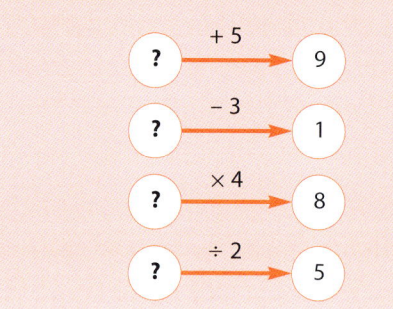

  ? →(+5)→ 9
  ? →(−3)→ 1
  ? →(×4)→ 8
  ? →(÷2)→ 5

A1 Write down the missing number at the end of each arrow diagram.

(a) 3 →(+8)→ ? (b) 12 →(−5)→ ?

(c) 6 →(×2)→ ? (d) 9 →(÷3)→ ?

A2 Find the missing number at the beginning of each arrow diagram.

(a) ? →(+2)→ 6 (b) ? →(×2)→ 10

(c) ? →(−5)→ 3 (d) ? →(÷5)→ 3

A3 Find the missing number in each arrow diagram.

(a) 12 →(+?)→ 21 (b) 7 →(×?)→ 21

B Think of a number

level 3

B1 "I think of a number." "I add 5." "The result is 9."

What number did she think of?

B2 "I think of a number." "I multiply by 4." "The result is 12."

What number did he think of?

B3 Solve each of these.

(a) I think of a number.
I add 3.
The result is 5.
What number did I think of?

(b) I think of a number.
I double it.
The result is 6.
What number did I think of?

(c) I think of a number.
I take away 1.
The result is 3.
What number did I think of?

(d) I think of a number.
I multiply by 10.
The result is 70.
What number did I think of?

(e) I think of a number.
I subtract 4.
The result is 11.
What number did I think of?

(f) I think of a number.
I divide by 2.
The result is 7.
What number did I think of?

B4 Solve each of these.

(a) I think of a number.
I multiply by 8.
The result is 72.
What number did I think of?

(b) I think of a number.
I subtract 20.
The result is 5.
What number did I think of?

(c) I think of a number.
I divide by 5.
The result is 8.
What number did I think of?

28 Starting equations

C Number puzzles level 4

- Find the missing number in each of these.

  | ▼ + 1 = 5 | ● × 3 = 12 | ■ − 4 = 3 | ★ ÷ 3 = 2 |

C1 Find the missing number in each of these.
(a) ♥ + 3 = 9
(b) ☆ − 2 = 5
(c) ❋ × 2 = 4
(d) ◆ − 7 = 3
(e) ✜ × 3 = 15
(f) ✪ ÷ 2 = 5
(g) ✹ ÷ 4 = 2
(h) ✿ − 6 = 14
(i) ✱ × 5 = 25

- Find the missing number in each of these.

  | 3 + ♥ = 10 | 6 × ★ = 12 | 5 − ◆ = 4 | 16 ÷ ▶ = 8 |

C2 Find the missing number in each of these.
(a) 11 + ■ = 20
(b) 7 × ✿ = 14
(c) 10 − ❋ = 8
(d) 15 − ☆ = 11
(e) 6 ÷ ✜ = 2
(f) 30 ÷ ◆ = 3

C3 Find the missing number in each of these.
(a) ☆ × 9 = 27
(b) 4 × ▲ = 20
(c) ✚ − 9 = 2
(d) 10 − ✜ = 3
(e) 30 ÷ ✹ = 15
(f) ✱ ÷ 6 = 4

C4 Find the missing number in each of these.
(a) ▼ + ▼ = 8
(b) ■ + ■ + ■ = 6
(c) ☺ + ☺ + ☺ = 90

D Using letters

In mathematics, a letter is often used to stand for a missing number.
- Find the missing number in each of these.

  | $n + 5 = 11$ | $3 \times a = 24$ | $x − 4 = 11$ | $10 ÷ b = 5$ |

D1 Find the missing number in each of these.
(a) $n + 1 = 5$
(b) $n + 3 = 9$
(c) $4 + n = 6$
(d) $8 + n = 14$

D2 Find the missing number in each of these.
(a) $p \times 3 = 18$
(b) $6 \times p = 12$
(c) $p \times 2 = 24$
(d) $8 \times p = 64$

D3 Find the missing number in each of these.
(a) $x − 7 = 1$
(b) $x − 4 = 3$
(c) $7 − x = 1$
(d) $4 − x = 3$

D4 Find the missing number in each of these.

(a) $y \div 3 = 6$ (b) $36 \div y = 12$ (c) $y \div 4 = 4$ (d) $20 \div y = 2$

A statement such as $n + 3 = 9$ is an example of an **equation**.
To **solve** this equation is to find the missing number that n stands for (called the **solution**).
$6 + 3 = 9$, so the missing number (or solution) is 6.
We can write this solution as $n = 6$.

D5 Solve each of these.
Write each solution as '$n = ...$'.

(a) $n + 6 = 8$ (b) $12 - n = 11$ (c) $7 + n = 20$ (d) $n \times 6 = 42$
(e) $27 \div n = 9$ (f) $9 \times n = 9$ (g) $n - 10 = 25$ (h) $n \div 5 = 11$

Shorthand is sometimes used for multiplications: for example, $3n$ is shorthand for $3 \times n$.
So $3n = 30$ means the same as $3 \times n = 30$.

D6 Solve each of these.

(a) $3n = 30$ (b) $2x = 24$ (c) $10y = 40$ (d) $7k = 35$

An equation such as $n + 4 = 9$ can also be written as $9 = n + 4$.

D7 Solve each of these.

(a) $9 = n + 4$ (b) $23 = 16 + m$ (c) $32 = 4p$ (d) $9 = 10 - q$

D8 Solve each of these.

(a) $n + 15 = 20$ (b) $3 \times n = 27$ (c) $n - 7 = 21$ (d) $n \div 5 = 10$
(e) $12 = x + 5$ (f) $x \times 7 = 49$ (g) $17 + x = 24$ (h) $4x = 20$
(i) $32 - a = 10$ (j) $24 \div a = 3$ (k) $14 = 2a$ (l) $18 = a - 2$

D9 Use sheets FT–19 and FT–20 and play 'The equation game' with one other person.
You solve equations, cover up the solutions with counters and try to make a line of four.

Test yourself

T1 Find the missing numbers.

(a) $36 - \blacklozenge = 20$ (b) $24 \div \clubsuit = 4$ (c) $\maltese + 18 = 26$ (d) $8 \times \circledast = 40$ OCR

T2 Solve (a) $a + 2 = 7$ (b) $10 - b = 8$ (c) $3c = 12$ OCR

T3 Solve (a) $x - 3 = 17$ (b) $3x = 18$ (c) $10 = x + 2$ OCR

29 Finding your way

You will revise using simple grid references and the points of the compass, estimating distances from a scale, giving directions, and distinguishing between a clockwise and anticlockwise turn.

A Using a town plan

level 4

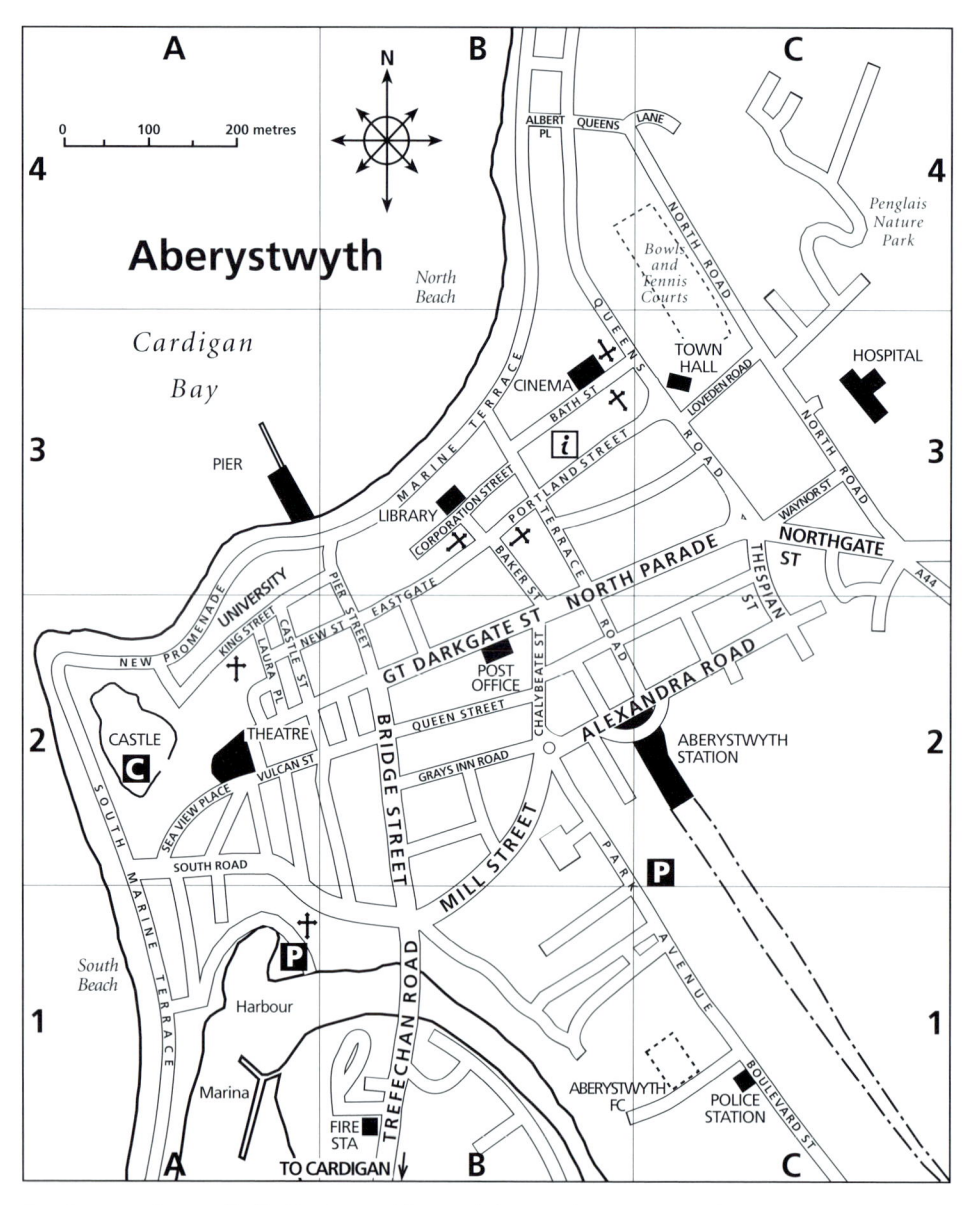

A1 (a) The library is in grid square B3. Which street is it on?

(b) On which street is the police station in grid square C1?

(c) In which grid square is

    (i) the pier   **(ii)** the railway station   **(iii)** the marina   **(iv)** the hospital

A2 You are walking along South Road (A2) away from the sea. Would you turn left or right to go into Bridge Street?

A3 Would you turn left or right if you were

(a) walking along Pier Street (B3) away from the pier and turned down Great Darkgate Street

(b) driving along Alexandra Road (C2) toward the roundabout and turned into Terrace Road

(c) walking down Marine Terrace (B3) toward the pier and turned into Terrace Road

A4 If you were standing on the castle walls (A2) at point C and looked toward these places, in what compass direction would you be facing?

(a) The station (C2)   **(b)** The pier   **(c)** Aberystwyth Football Club (C1)

A5 In what compass direction would you be travelling if you were

(a) driving along Trefechan Road (B1) from the town to the fire station

(b) walking along Terrace Road (B3) toward the sea

(c) driving along Portland Street (B3) away from the town hall

A6 Where would these directions take you, starting at the town hall (C3)?

    Head south-east along Queens Road.
    At the end of Queens Road turn right and take the first turning on the left.
    Go to the end of this road and go straight across.
    You'll see it straight in front of you.

A7 You are standing in the middle of the roundabout where Mill Street meets Alexandra Road (B2). You face towards Chalybeate Street then turn slowly clockwise. Give the names, in order, of the other four roads you will look down as you do this.

A8 Give careful directions to someone who wants to walk

(a) from the post office (B2) to the railway station (C2)

(b) from the cinema (B3) to the hospital (C3)

(c) from the railway station (C2) to the fire station (B1)

A9 Use the scale and a piece of paper to estimate how far it is to walk from

(a) the railway station (C2) to the cinema (B3)

(b) the theatre (A2) to the hospital building (C3)

Test yourself

Here is a sketch map of the town of Ringsford.

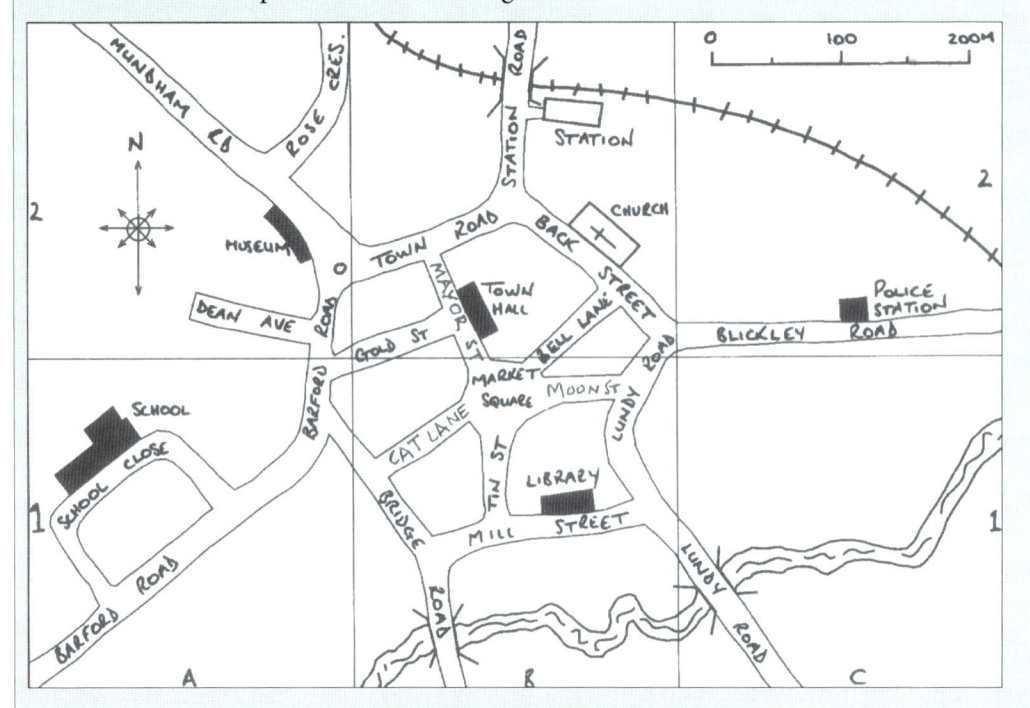

T1 In which square will you find (a) the police station (b) the school

T2 You are walking along Gold Street toward the town hall. Would you turn left or right at the end to go to the market square?

T3 In what compass direction would you be going if you were
 (a) walking out of town on Blickley Road (C2) toward the police station
 (b) walking along Barford Road (A1) away from the town centre

T4 You are standing somewhere in the town. By turning slowly you look down these streets in order: Tin Street, Moon Street, Bell Lane, Mayor Street and Cat Lane. Where are you standing and which way are you turning?

T5 Where would these directions take you?
 Come out of the town hall and go left to the market square.
 Leave the market square going south.
 Turn left and you will find this on the left-hand side of the road you are on.

T6 Give directions to someone from outside the library (B1) to the school (A1).

T7 Use the scale and a piece of paper to estimate the distance from the museum (A2) to the police station (C2).

Review 5

1 (a) Measure the edges of this triangle and find its perimeter.
  (b) Is the triangle scalene, isosceles or equilateral?
  (c) (i) On centimetre squared paper, draw an enlargement of the triangle with scale factor 2.
   (ii) What is the perimeter of the enlarged triangle?
  (d) Find the area of the larger triangle.

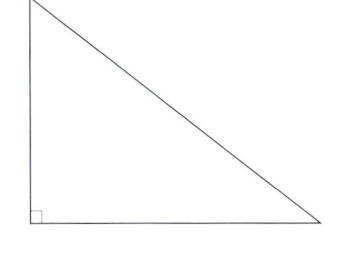

2 At noon the temperature in Kelly's garden was 8 °C.
By midnight the temperature had fallen by 10 degrees.
What was the temperature at midnight?

3 Find the missing number in this arrow diagram. ? —×5→ 10

4 Write this list of temperatures in order, starting with the lowest.
   1 °C   ⁻4 °C   ⁻2 °C   0 °C   3 °C

5 This is a sketch map of a park.

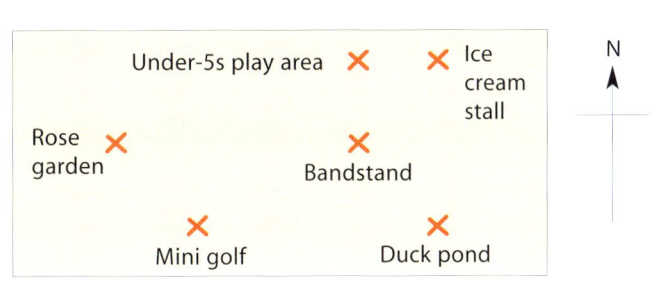

  (a) Rikki walks north from the duckpond. Which of the marked places does he get to?
  (b) Mike walks south-east from the bandstand. Which of the marked places does he get to?
  (c) Penny walks from the mini golf to the rose garden. Which compass direction is this?

6 Solve each of these equations.
  (a) $n + 5 = 8$ **(b)** $x - 3 = 6$ **(c)** $2m = 18$ **(d)** $y \div 3 = 5$

7 Two friends records how many cups of coffee they drink each day at work for a week.
   Alex   3   5   4   2   4
   Becky 3   7   1   4   5

  (a) Find the mean number of cups for each person.
  (b) Find the range for each person.

30 Volume

This work will help you find the volume in cubic centimetres of a solid.

A Counting cubes

A cube this size is called a centimetre cube.
It has a volume of 1 cubic centimetre (1 cm³).

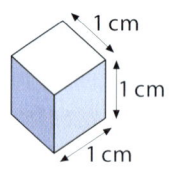

A1 These shapes are made out of centimetre cubes.
Find the volume of each shape. Give each answer with cm³ as the units.

(a) 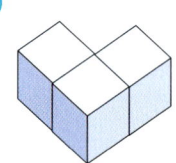　(b) 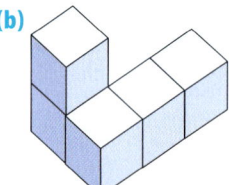　(c) 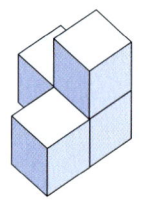　(d)

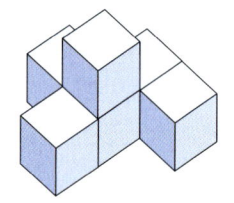

A2 This prism has been made out of centimetre cubes.
In this picture some cubes are hidden behind other cubes.
The prism has a volume of 15 cm³.
How many cubes are completely hidden?

Here is a rectangular layer of centimetre cubes.

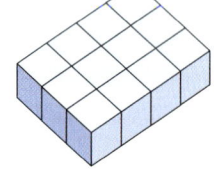

You can think of it as split into rows.
Now you can work out the number of cubes without counting each one.

There are 3 rows. Each row has 4 cubes.
So there are 3 × 4 cubes in the layer.
So the volume of the layer is 12 cm³.

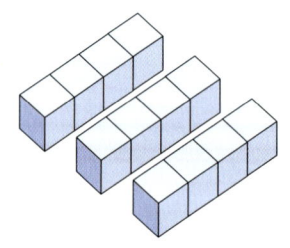

A3 For each of these rectangular layers of cubes, write a multiplication to find its volume.

(a) 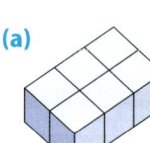　(b)

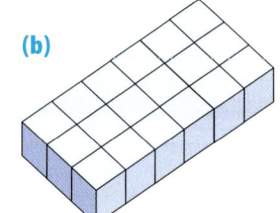

A4 Find the volume of each of these rectangular layers of cubes.

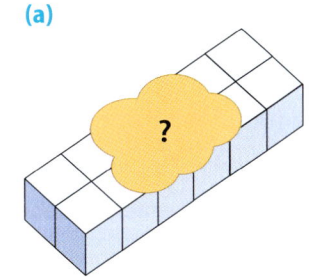

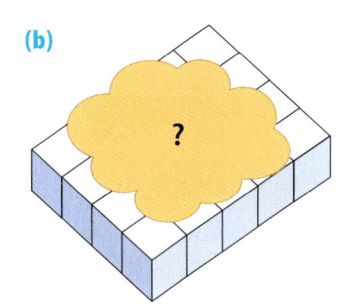

(a)

(b)

A5 This pyramid is made out of centimetre cubes. Each layer is a square.

This layer is a 2 cm by 2 cm square.
This layer is 3 cm by 3 cm.
This is 4 cm by 4 cm.

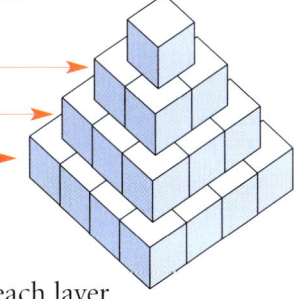

(a) Write down the number of cubes in each layer. (Don't forget the single cube on top.)

(b) Now work out the volume of the whole pyramid in cm³.

A6 Some centimetre cubes have been piled up in layers. Each layer is a rectangle. We call the complete shape a **cuboid**.

(a) How many cubes are there on the top layer?
(b) How many layers are there?
(c) Multiply to find the volume of the whole cuboid in cm³.

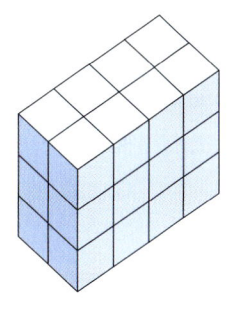

Test yourself

T1 These shapes are made out of centimetre cubes. Find the volume of each one.

(a)

(b)

(c)

31 Evaluating expressions

You will revise how letters can stand for numbers in expressions and formulas.

This work will help you substitute in simple expressions and formulas.

You need sheets FT–21 and FT–22, and some counters.

A Simple substitution

There is an expression in each box below.

3n means 3×n.

A $n + 5$　　**B** $n - 2$　　**C** $3n$　　**D** $n \div 2$

- Find the value of each expression when $n = 6$.
- Evaluate each expression when $n = 0$.

A1 What is the value of each expression when $n = 4$?
  (a) $n + 2$　　(b) $n + 5$　　(c) $n - 3$　　(d) $n - 4$

A2 What is the value of each expression when $n = 3$?
  (a) $2n$　　(b) $5n$　　(c) $4n$　　(d) $10n$

A3 Evaluate each expression when $n = 12$.
  (a) $n \div 2$　　(b) $n \div 3$　　(c) $n \div 4$　　(d) $n \div 6$

A4 Evaluate each expression when $x = 15$.
  (a) $x - 9$　　(b) $2x$　　(c) $x + 5$　　(d) $x \div 5$

A5 Copy and complete this table to show the value of each expression for some different values of n.

	$n = 2$	$n = 4$	$n = 6$	$n = 8$
$n + 3$			9	
$n - 2$				
$3n$				
$n \div 2$	1			

*A6 Each expression in the diagram stands for the length of a side in centimetres.
  (a) (i) Work out the length of each side when $x = 3$. Sketch the triangle.
  　　(ii) What is the perimeter of this triangle?
  (b) What is the perimeter of the triangle when $x = 5$?

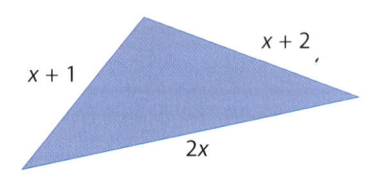

Triangle with sides $x + 1$, $x + 2$, $2x$.

B Rules for calculation

- Can you work these out?

  A $(10 - 2) \times 3$
  B $5 \times (3 - 1)$
  C $10 + 2 \times 3$
  D $5 \times 3 - 1$
  E $6 \div 2 + 4$
  F $(6 + 4) \div 2$
  G $6 + 4 \div 2$
  H $6 - 4 \div 2$
  I $(6 + 4) \div (3 - 1)$

  Evaluate expressions in brackets **first**.

  To calculate with +, −, × and ÷, multiply or divide **before** you add or subtract.

B1 Work out each of these.
- (a) $(6 + 3) \times 2$
- (b) $3 \times (2 + 5)$
- (c) $2 \times (10 - 4)$
- (d) $2 \times 6 - 5$
- (e) $4 \times 5 - 3$
- (f) $10 + 3 \times 3$
- (g) $2 + 7 \times 4$
- (h) $15 - 2 \times 5$
- (i) $20 - 3 \times 6$

B2 Work out each of these.
- (a) $12 \div 3 + 6$
- (b) $(8 + 2) \div 5$
- (c) $8 + 16 \div 4$
- (d) $15 \div 5 - 1$
- (e) $(15 - 1) \div 7$
- (f) $16 - 8 \div 4$

B3 Work out each of these.
- (a) $(1 + 3) \times (2 + 5)$
- (b) $(10 - 5) \times (6 - 2)$
- (c) $(25 - 1) \div (2 + 2)$
- (d) $(9 + 21) \div (8 - 2)$
- (e) $3 \times (7 + 1) + 5$
- (f) $2 \times (18 - 4) - 7$

$\frac{6}{2}$ is another way to write $6 \div 2$.

$\frac{6 + 4}{2}$ is another way to write $(6 + 4) \div 2$.

B4 Work out each of these.
- (a) $\frac{14}{2} + 5$
- (b) $1 + \frac{15}{5}$
- (c) $\frac{3 + 5}{4}$
- (d) $\frac{20 - 10}{2}$
- (e) $\frac{20}{2} - 3$
- (f) $10 - \frac{6}{2}$

B5 Work out each of these.
- (a) $\frac{14 + 4}{4 + 5}$
- (b) $\frac{18 - 3}{1 + 4}$
- (c) $\frac{22 - 8}{11 - 9}$

B6 Find the missing number in each of these calculations.
- (a) $5 \times 6 + \blacksquare = 40$
- (b) $(6 + 2) \times \blacksquare = 24$
- (c) $(\blacksquare - 5) \times 3 = 15$
- (d) $\frac{10 - \blacksquare}{4} = 2$
- (e) $\frac{\blacksquare}{3} + 9 = 14$
- (f) $20 - \frac{10}{\blacksquare} = 15$

31 Evaluating expressions

C Substituting into linear expressions

Examples

Find the value of $3(n-1)$ when $n = 5$.

$3(n-1)$ means $3 \times (n-1)$.

$$3(n-1) = 3 \times (5-1)$$
$$= 3 \times 4$$
$$= 12$$

Find the value of $\dfrac{x-5}{2}$ when $x = 11$.

$$\dfrac{x-5}{2} = \dfrac{11-5}{2}$$
$$= \dfrac{6}{2}$$
$$= 3$$

Find the value of $\dfrac{a}{3} + 9$ when $a = 12$.

$$\dfrac{a}{3} + 9 = \dfrac{12}{3} + 9$$
$$= 4 + 9$$
$$= 13$$

C1 What is the value of each expression when $a = 5$?
 (a) $2a - 1$ (b) $3a + 5$ (c) $2(a - 2)$ (d) $3(a + 1)$

C2 Evaluate each expression when $n = 9$.
 (a) $3n + 1$ (b) $5n - 6$ (c) $3(n + 1)$ (d) $5(n - 6)$

C3 What is the value of each expression when $x = 10$?
 (a) $\dfrac{x}{5} + 2$ (b) $\dfrac{x+2}{3}$ (c) $\dfrac{x-2}{4}$ (d) $\dfrac{x}{2} - 3$

C4 Evaluate each expression when $k = 12$.
 (a) $\dfrac{k}{6} - 1$ (b) $\dfrac{k+6}{9}$ (c) $\dfrac{k-8}{2}$ (d) $\dfrac{k}{3} + 5$

C5 (a) Copy and complete this table to show the value of each expression for some different values of n.

	$n = 3$	$n = 6$	$n = 9$	$n = 12$
$3n + 4$			31	
$5n - 2$				
$2(n - 1)$				
$4(1 + n)$	16			
$\dfrac{n+6}{3}$				
$\dfrac{n}{3} - 1$				

(b) (i) Which expression has a value of 40 when $n = 9$?
  (ii) Which expressions have the same value when $n = 3$?
  (iii) Which expressions have the same value when $n = 6$?
  (iv) Which expression has the greatest value when $n = 12$?
  (v) Which expression has the smallest value when $n = 9$?

*C6 Each expression in the diagram stands for the length of a side in centimetres.

(a) (i) Work out the length of each side when $x = 2$. Draw the rectangle.

(ii) What is the area of this rectangle?

(b) What is the area of the rectangle when $x = 10$?

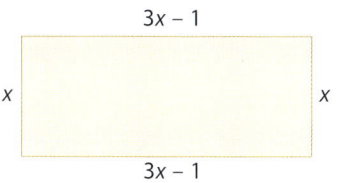

Sides labelled: $3x - 1$ (top and bottom), x (left and right).

Link up four a game for two players

What you need

- You need a dice.
  You each need about 12 counters of your own colour.
  You need a copy of the game board shown below (sheet FT–21).
  You also need a set of game cards (sheet FT–22).

18	3	10	2	6
8	9	6	7	4
13	4	7	5	12
7	5	8	3	9
14	6	11	1	15

Cards shown: $n + 3$, $n + 13$, $2(n + \ldots)$, $\dfrac{n}{3} + 5$

Before you start

- Deal five cards to each player (you do not need to keep them hidden).

When it is your turn

- Roll the dice.
- The number on the dice is the value of n for your turn.
- Find the value of any one of your expressions. Cover this number on the board with one of your counters.
- Discard the card you used and pick another.
- If none of your expressions gives a number on the board, you do not cover a number.
- You can only use one card on your turn.

The winner

- The winner is the first player to cover four numbers in a row (across, down or diagonally).

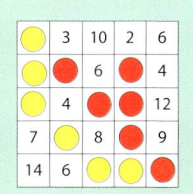

Red wins here.

31 Evaluating expressions

D Formulas in words

D1 Some undertakers use the rule that the length of a body is about three times the circumference of the head.

We can write this rule as

length = 3 × circumference

(a) Use this rule to estimate the length of a body whose head has a circumference of 50 cm.

(b) Use this rule to estimate the length of a body whose head has a circumference of 54 cm.

D2 Superstore Systems make CD racks.
They make racks to fit any number of CDs.

The base of each rack is 60 mm high.
Each CD space adds another 14 mm to the height of the rack.

The rule for the height of the rack is

height in mm = 14 × number of CDs + 60

Work out the height of a rack that holds

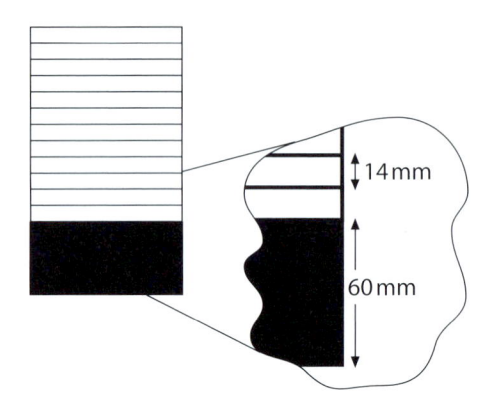

(a) 10 CDs (b) 30 CDs
(c) 50 CDs (d) 80 CDs

D3 Superstore Systems also make DVD racks.

The rule for the height of the rack is

height in mm = 20 × number of DVDs + 40

Work out the height of a rack that holds

(a) 10 DVDs (b) 15 DVDs
(c) 20 DVDs (d) 30 DVDs

D4 There is a rule for finding how far away thunderstorms are when you know how many seconds pass between seeing the lightning and hearing the thunder.

$$\text{distance in miles} = \frac{\text{time in seconds}}{5}$$

(a) Work out how far away the storm is if it takes 40 seconds to hear the thunder.

(b) How far away is the storm if you count 15 seconds between the thunder and lightning?

D5 The Rugged Walk outdoor centre organises walking trips.

They use this rule to work out the number of maps to take for groups of people.

$$\text{number of maps} = \frac{\text{number of people}}{2} + 1$$

Work out how many maps they would take for a group of

(a) 10 people (b) 16 people (c) 30 people

D6 Pru makes dresses in a factory.
This formula is used to calculate her weekly wage.

weekly wage (£) = 3 × number of dresses + 30

Work out her weekly wage for

(a) 30 dresses (b) 60 dresses (c) 80 dresses

D7 The number of points a soccer team has may be worked out using this formula.

number of points = 3 × number of wins + number of draws

Last season, Manchester United had 28 wins and 7 draws.
Work out the number of points they had.

Edexcel

D8 Some people work at the Pizza Parlour.
The formula used for their wages is

weekly wage = number of hours × hourly rate

(a) Sue washes dishes for 12 hours at an hourly rate of £4.50.
How much does she earn?

(b) Fran makes pizzas at an hourly rate of £6.00 for 28 hours.
How much does she earn?

D9 Cheryl was working out the cost of hiring a van for a day.

First of all she worked out the mileage cost.
She used the formula

mileage cost = mileage rate × number of miles travelled

The mileage rate was 8 pence per mile.
Cheryl travelled 280 miles.

(a) Work out the mileage cost in pounds.

Cheryl worked out the total cost by using the formula

total hire cost = (basic hire cost + mileage cost) × 1.2

The basic hire cost was £45.

(b) Work out the total hire cost.

Edexcel

E Formulas without words

Kim uses this rule to work out how many sandwiches to make for a party.

$S = 4p + 10$

S is the number of sandwiches and p is the number of people at the party.

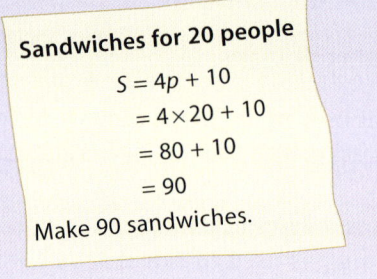

Sandwiches for 20 people
$S = 4p + 10$
$= 4 \times 20 + 10$
$= 80 + 10$
$= 90$
Make 90 sandwiches.

E1 A clothing shop uses the following rule to change British dress sizes to American ones.

$A = B - 4$

A is the American and B is the British dress size.

(a) Doreen wears a British dress size of 14.
What is her American dress size?

(b) Change the British size of 18 to the American size.

E2 The rule to work out the perimeter of a regular pentagon is

$P = 5s$

P is the perimeter of the pentagon and s is the length of one side.

Calculate the perimeter of a regular pentagon where one side measures

(a) 2 cm   (b) 7 cm   (c) 8 mm

E3 Peter uses this rule to work out how much mince to use for a cottage pie.

$W = 100n + 80$

W is the weight of mince in grams and n is the number of people.

How much mince would Peter buy to make a pie for

(a) 4 people   (b) 6 people   (c) 10 people

E4 Craig's band has five members.
After each performance, the whole band is paid.
They use this rule to work out how much each band member gets paid.

$M = \dfrac{F - 50}{5}$

M is the amount of money each band member gets and F is the total the band gets for a performance.

(a) One Saturday, the band was paid £225 for a night at Jo Jo's.
How much did each band member get?

(b) How much does each band member get if the total earned is £450?

Test yourself

T1 Evaluate each of these.

(a) $4 \times (5 - 2)$ (b) $\dfrac{9 + 12}{3}$ (c) $20 - \dfrac{9}{3}$

T2 Work these out.

(a) $8 - 3 \times 2$ (b) $(30 + 20) \div (5 - 3)$ OCR

T3 What is the value of each expression when $n = 8$?

(a) $2n - 10$ (b) $5(n + 1)$ (c) $\dfrac{n}{4} + 12$ (d) $\dfrac{n + 2}{5}$

T4 Evaluate $\dfrac{12 + x}{2}$ when

(a) $x = 2$ (b) $x = 8$ (c) $x = 10$ (d) $x = 0$

T5 (a) Pro-print uses this formula to work out the price, in pounds, of posters.

> Multiply the number of posters by 3, then add 25

Work out the price of 15 posters from Pro-print.

(b) Fasta-print uses this formula to work out the price, in pounds, of posters.

$P = 4 \times n + 18$

P is the price in pounds.
n is the number of posters.

Work out the price of 12 posters from Fasta-print. OCR

T6 The Rugged Walk outdoor centre organises walking trips.

They use this rule to work out the number of sandwiches to take for groups of people.

$S = 3p + 5$

S is the number of sandwiches and p is the number of people.

Work out how many sandwiches they would take for a group of

(a) 5 people (b) 10 people (c) 16 people

31 Evaluating expressions 143

32 Estimating and calculating with money

This work will help you

- solve problems involving money without a calculator
- estimate answers to problems involving money
- solve problems using a calculator

A Solving problems without a calculator   level 4

Mr and Mrs Field and their three children visit the zoo.
What is the total cost of their tickets?

Total cost	£6.00
	£6.00
	£3.50
	£3.50
	+ £3.50
	£22.50

Adults   £6.00 × 2 = £12.00
Children   £3.50 × 3 = £10.50
Total   £12.00 + £10.50 = £22.50

Entry
Adults £6·00
Children £3·50
Seniors £3·00

- How would you work out the total cost for one senior and three children?

A1 Work out the total cost of the following tickets for the zoo.
Show clearly how you get your answer each time.

(a) One adult and two children
(b) One adult, one child and one senior
(c) Two seniors and two children
(d) Three adults and four children

A2 Mr Jackson takes his three children to the zoo.
He pays for the tickets with a £20 note.
How much change does he get?

A3 The zoo has a special family ticket for £17.50.
Keith and Laura are taking their three children to the zoo.

How much cheaper is it to buy a family ticket instead of tickets for two adults and three children?

Special offer
FAMILY TICKET
only £17.50
(up to 2 adults + 3 children)

A4 Is it cheaper to buy a family ticket instead of tickets for two adults and two children?
If so, how much cheaper is it?

A5 Animal food costs £1.25 per bag.

(a) Simon buys 2 bags of animal food.
How much does he pay?

(b) Rani buys 3 bags of animal food and pays with a £5 note.
How much change does she get?

(c) Petra has £4.00.
How many bags of animal food can she buy?

A6 Postcards cost 40p each.
Erin has £1.50 to spend.

(a) How many postcards can she buy?

(b) How much money will she have left?

A7 There is a cable car ride at the zoo.
Each car holds up to 4 people.

How many cars are needed for

(a) a group of 8 people   (b) a group of 14 people

B Estimating answers   level 4

Sometimes it is useful to make a rough estimate of an answer.

Menu
Coffee	£1.99
Tea	£1.79
Juice	£0.95
Muffin	£0.99
Sandwich	£2.90

I want 2 coffees and a juice. I've only got £5. Is that enough?

Roughly how much will one juice, one sandwich and one muffin cost me?

I've got £10. If I buy 2 teas and 2 sandwiches, will I have enough money left for a muffin?

I've got £4. I want a drink and a snack. What could I buy?

How many coffees can I buy with £8?

B1 (a) James buys 6 DVDs.
Roughly how much do they cost?

SALE
All DVDs now
£5.99

(b) Lena has £15.
How many DVDs can she buy?

B2 (a) Trisha has £3.
How many oranges can she buy?

ORANGES 39p each

MELONS £1.89 each

(b) Marlon buys a melon and an orange.
Roughly how much do they cost altogether?

(c) Zoe has £5. She wants to buy 2 melons and 3 oranges.
Has she got enough money?

B3 (a) Colin pays for two adults to go skating.
Roughly how much does it cost him?

Skating prices
Adults   £5.95
Children £4.95

(b) Kate buys tickets for a group of 10 children.
Does it cost her more or less than £50?
Explain how you know.

(c) Manny buys tickets for two adults and two children.
Does it cost him more or less than £20?
Explain how you know.

C Using a calculator   level 4

I buy 8 pens at £1.45 each.
How much do they cost altogether?

Do 1.45 × 8 on your calculator

11.6

- How would you write this answer?

- How much change would you get if you paid with £15?

Use your calculator to work these out.

C1 Work out the cost of

(a) 14 notebooks at £1.25 each
(b) 6 calculators at £3.80 each

C2 Pencils are sold in packs of 12.
Phil needs 30 pencils.

(a) How many packs does he need to buy?

(b) A pack of pencils costs £2.79.
How much does Phil spend on the pencils?

C3 Chris buys a Swiss ball and a yoga DVD.
How much does he pay altogether?

C4 Robin buys a yoga mat and a yoga block.
   (a) How much do they cost altogether?
   (b) He pays with £30.
       How much change does he get?

C5 Jeanette wants to buy some Swiss balls.
She has £50.
How many Swiss balls can she buy?

C6 A yoga starter pack contains a yoga mat, block and DVD.
The starter pack costs £35.

How much money do you save if you buy the starter pack instead of buying the items individually?

C7 (a) Maureen charges £7.50 for her yoga class.
       How much money does she collect if 24 people attend the class?
   (b) Maureen pays £45 to hire the room.
       How much money does she have left after paying for the room?

Swiss ball £16.50
Yoga mat £19.99
DVD £12.50
Yoga block £3.95

Test yourself

T1 In a book shop, all sale books cost £2.99.
Roughly how much will 4 sale books cost?

T2 Jim went to the bowling alley.
It cost him £1.50 to hire shoes and £2.50 per game.
Jim paid £9 in total.
How many games did he pay for?
OCR

T3 A garden centre has tomato plants for sale.

Work out the cheapest price for 24 tomato plants.

Tomato plants
40 pence each
or
£5 for a box of 20

AQA

T4 Tony is buying some CDs.
Each CD costs £8.95.
   (a) How many CDs can Tony buy for £40?
   (b) How much money does he have left?
AQA

33 Capacity

This work will help you with litres and millilitres.

A Litres, millilitres and other metric units

Imagine a tiny container in the shape of a centimetre cube that is full of liquid.

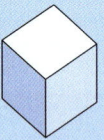

The volume of liquid in this tiny container would be 1 **millilitre** (1 ml).
The maximum volume of liquid that a container can hold is called its **capacity**.
What do you think could be the capacity in millilitres of
- a teaspoon?
- an eggcup?
- a mug?

A **litre** is a thousand millilitres.
What do you think could be the capacity in litres of
- a bucket?
- a bath?
- a swimming pool?

A1 Which of these could be the capacity of a teacup?
 25 ml   250 ml   2.5 litres   25 litres

A2 Which of these could be the capacity of a soup bowl?
 5 ml   50 ml   500 ml   5 litres

A3 Which of these could be the capacity of a kettle?
 17 ml   170 ml   1.7 litres   17 litres

A4 Which of these could be the capacity of a kitchen sink?
 2 ml   20 ml   2 litres   20 litres

A5 How many litres do these containers hold?

(a) 2000 ml   (b) 1500 ml   (c) 50 000 ml

A6 Fiona uses 400 ml of milk from a full container that holds 1 litre.
How much milk is left?

A7 How many millilitres are in $\frac{3}{4}$ of a litre?

A8 Write each of these as a decimal of a litre.

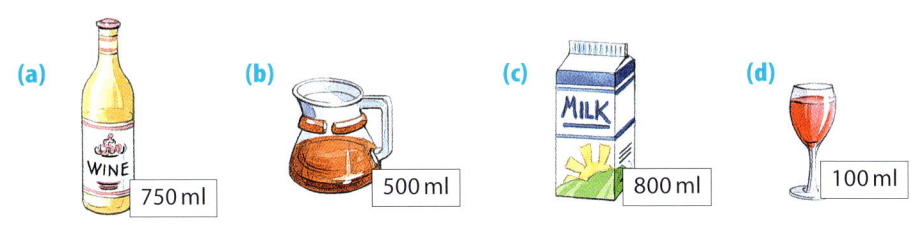

(a) 750 ml (b) 500 ml (c) 800 ml (d) 100 ml

A9 How many millilitres are there in each of these?
(a) 1.5 litres (b) 5.2 litres (c) 10 litres (d) 0.4 litre

A10 How many litres are there in each of these?
(a) 3000 ml (b) 4500 ml (c) 1650 ml (d) 6200 ml

A11 Liquid medicine is usually given in 5 ml spoonfuls.
How many 5 ml spoonfuls are in a bottle that holds 0.5 litre of medicine?

A12 Write these volumes in order, smallest first.
750 ml $\frac{1}{2}$ litre   1.2 litres   50 ml   1500 ml

Millimetres, centimetres, metres, kilometres, grams, kilograms, millilitres and litres are all **metric units**.

A13 Which metric unit completes each statement?
(a) The height of a fridge is 1.5 ___ .
(b) A vacuum flask holds 700 ___ of liquid.
(c) The length of a pencil case is 17.8 ___ .
(d) The weight of a chicken is 3 ___ .

Test yourself

T1 Put these volumes in order, **largest** first.
500 ml   4 litres   250 ml   1.5 litres   OCR

T2 Fiona uses 600 ml of olive oil from a full bottle that holds 1 litre.
How much olive oil is left?

T3 (a) How many litres are there in (i) 8000 ml (ii) 3500 ml
(b) How many millilitres are in $1\frac{1}{4}$ litres?

T4 Which metric unit completes each statement?
(a) A can of beans weighs 400 ___ .   (b) The diameter of a £1 coin is 22 ___ .

34 Drawing and using graphs

You should know how to
- plot coordinates
- work out expressions like $3 + 4n$ when $n = 2$

This work will help you draw and interpret straight-line graphs for real-life situations.

A Tables and graphs

A1 Aluna is filling a fish tank.

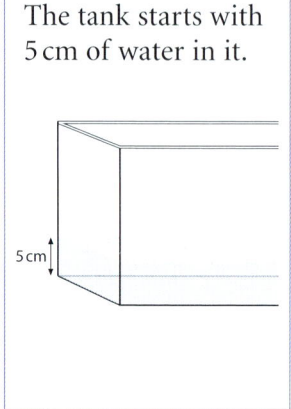

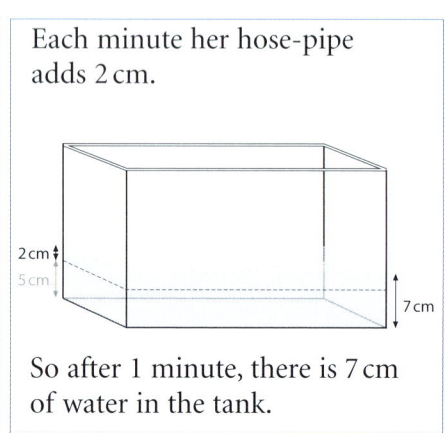

The tank starts with 5 cm of water in it.

Each minute her hose-pipe adds 2 cm.

So after 1 minute, there is 7 cm of water in the tank.

(a) How deep will the water be after 2 minutes?

(b) Copy and complete this table.

Time in minutes	0	1	2	3	4	5
Depth of water in cm	5	7				

(c) On graph paper draw and label axes like these. Then plot the points from your table (the first two are shown already plotted).

Go up to 20.

Stop at 7.

(d) Join the points you have plotted. Extend the line they make.

(e) Use your graph to say how deep the water will be after 7 minutes.

(f) How deep will the water be after $3\frac{1}{2}$ minutes?

(g) How many minutes will it take until the water is 10 cm deep?

(h) Aluna starts filling the tank at exactly quarter past 3. At what time will the water be 17 cm deep?

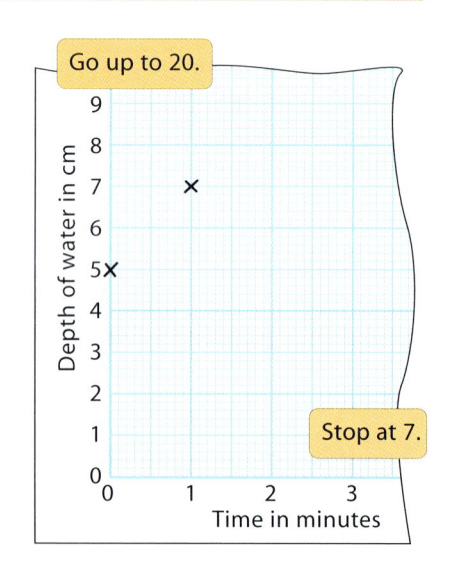

A2

This candle is 26 cm tall when it is first lit.

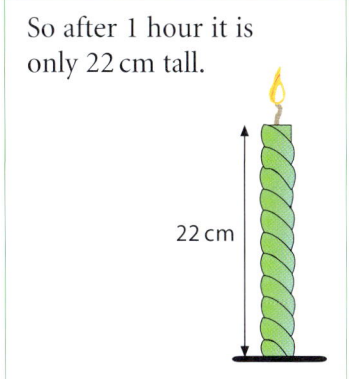

After it is lit, it gets shorter by 4 cm each hour.

So after 1 hour it is only 22 cm tall.

(a) How tall will the candle be after 2 hours?

(b) Copy and complete this table.

Time in hours	0	1	2	3	4
Height of candle in cm	26	22			

(c) On graph paper, draw and label axes like these. Then plot the points from your table.

(d) Join the points you have plotted. Extend the line they make.

(e) Use your graph to say how tall the candle will be after 6 hours.

(f) How tall will it be after $4\frac{1}{2}$ hours?

(g) How many hours will it take until the candle is 12 cm tall?

(h) How long will the candle last until it goes out?

(i) If the candle was lit at half past one, at what time will it go out?

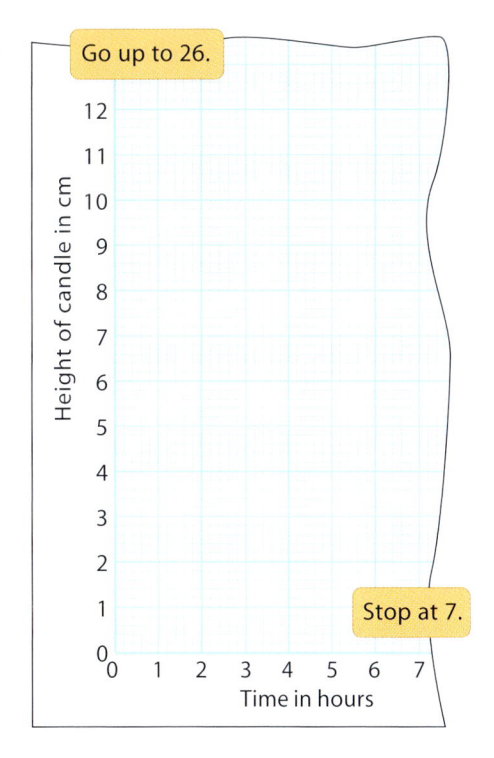

Go up to 26.

Stop at 7.

A3 Labib is slowly heating up a liquid in a science experiment.
The liquid starts at 20°C, and heats up by 8 degrees every 5 minutes.

(a) Copy and complete this table.

Time in minutes	0	5	10	15	20	25
Temperature in °C	20	28				

(b) Plot the points from your table on axes like the ones below.

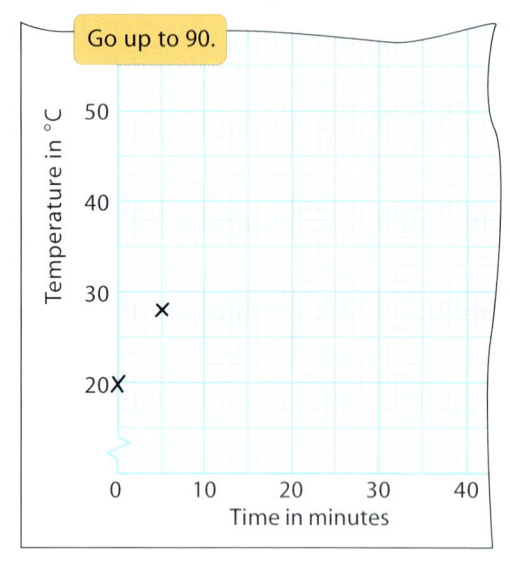

Go up to 90.

(c) Join your points and extend the line.
(d) Use the graph to find out what the temperature will be after 40 minutes.
(e) About how long will it take for the liquid to get to 30°C?
(f) Labib needs to get the liquid to 70°C.
He starts the experiment at a quarter past 4.
At about what time does the liquid get to 70°C?

A4 Chemico make anti-freeze for car radiators.

They test each batch of anti-freeze by cooling a sample.

At the start of the test, the anti-freeze is at 15 °C.
Then the temperature of the anti-freeze drops by 4 degrees each hour.

(a) What will the temperature of the anti-freeze be after 1 hour?

(b) After 5 hours, by how many degrees will the temperature have dropped?
What will the temperature of the anti-freeze be after 5 hours?

(c) Copy and complete this table.

Number of hours	0	1	2	3	4	5
Temperature in °C	15					

(d) Draw axes like these and plot the points from your table.
Join the points with a line, and extend it.

(e) Use your graph to find out what the temperature of the anti-freeze will be after $2\frac{1}{2}$ hours.

(f) What will the temperature be after 6 hours?

(g) How long will it take for the temperature of the anti-freeze to drop to −3 °C?

(h) If Chemico start testing the anti-freeze at 10 a.m. at what time will the temperature be −7 °C?

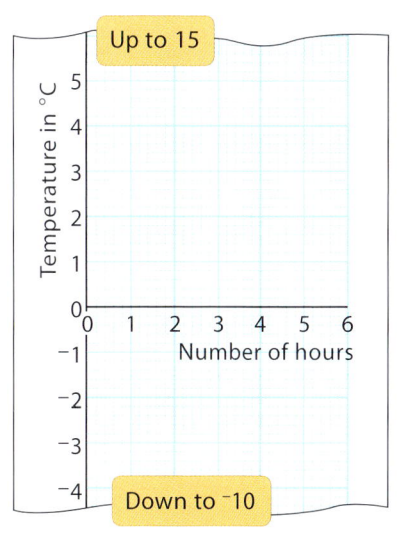

34 Drawing and using graphs 153

B Graphs and rules

B1 You can use this rule to change miles into kilometres.

number of kilometres = number of miles × 8 ÷ 5

(a) Use the rule to check that 10 miles is the same as 16 kilometres.

(b) Work out how many kilometres there are in

   (i) 20 miles   (ii) 30 miles   (iii) 50 miles

(c) Copy and complete this table.

Miles	0	10	20	30	40	50
Kilometres	0	16				

(d) Plot the points from your table using axes like these. Join the points with a line and extend it.

(e) About how many kilometres are there in

   (i) 42 miles   (ii) 55 miles

(f) About how many miles is

   (i) 20 kilometres   (ii) 75 kilometres

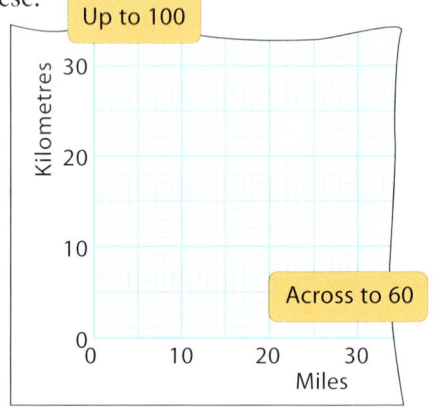

(g) On the road-signs below, the distances are in kilometres. Sketch each sign with the distances written in miles.

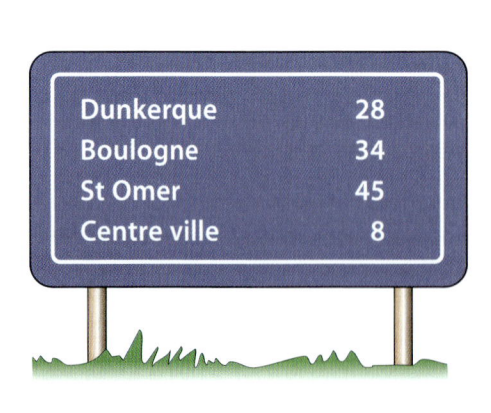

Dunkerque	28
Boulogne	34
St Omer	45
Centre ville	8

Paris	85
Versailles	78
Aéroport d'Orly	65
Fontainebleau	36
Gare TGV	15

B2 Z car taxis have meters in their taxis to work out the charge for each trip. The meters use the formula

$c = 4m + 1$

c is the charge in pounds; m is the number of miles the trip takes.

For example, if a customer travels 2 miles,

$c = 4 \times 2 + 1 = 9$

so the charge on the meter will be £9.00.

Z car taxis
2·0 miles
charge £ 9·00

(a) Copy and complete this table showing the charge for different lengths of journey.

Miles travelled (m)	1	2	3	4	5	6
Charge in £ (c)		9				

(b) Draw a graph to show the values in your table. Draw and label your axes like this.

Plot the points from your table and join them with a line.

Extend your line in both directions.

Label the line Z car taxis.

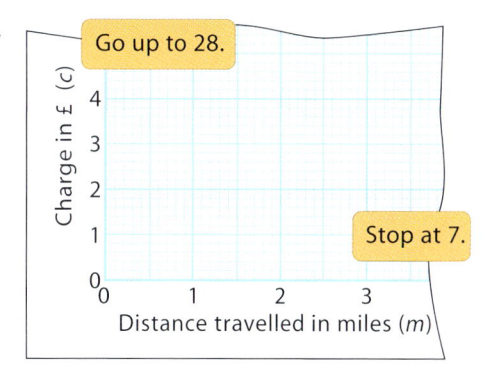

Go up to 28.
Stop at 7.

(c) Use your graph to find the missing numbers on these meters.

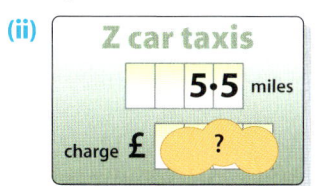

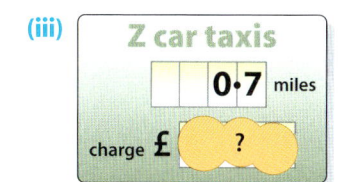

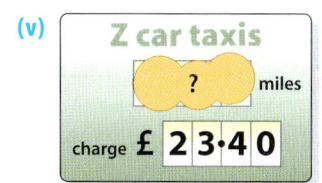

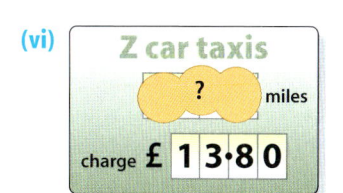

(i) Z car taxis — 1·8 miles, charge £ ?

(ii) Z car taxis — 5·5 miles, charge £ ?

(iii) Z car taxis — 0·7 miles, charge £ ?

(iv) Z car taxis — ? miles, charge £ 15·00

(v) Z car taxis — ? miles, charge £ 23·40

(vi) Z car taxis — ? miles, charge £ 13·80

34 Drawing and using graphs 155

For this question, you need the graph you drew in question B2.

B3 Aardvark taxis use a different formula.
Their formula is $c = 4.5 + 3m$.
So for a trip of 2 miles, $c = 4.5 + 3 \times 2 = 10.5$.
The charge will be £10.50.

(a) Copy and complete this table for Aardvark taxis.

Miles travelled (m)	1	2	3	4	5	6
Charge in £ (c)		10.50				

(b) Use the values in your table to plot a graph for the charges Aardvark taxis make.
Plot the graph on the same axes as for Z car taxis.
Label the line Aardvark taxis.

Use your graphs to answer these questions.

(c) (i) How much do Z cars charge for a trip of 1.5 miles?
   (ii) How much do Aardvark charge for the same distance?

(d) Which is cheaper, Z cars or Aardvark, for a trip of 1.5 miles?

(e) Which is cheaper, Z cars or Aardvark, for a trip of 4.5 miles?

(f) For one distance, Z cars and Aardvark charge the same.
What distance is that?

(g) Copy and complete this advice.

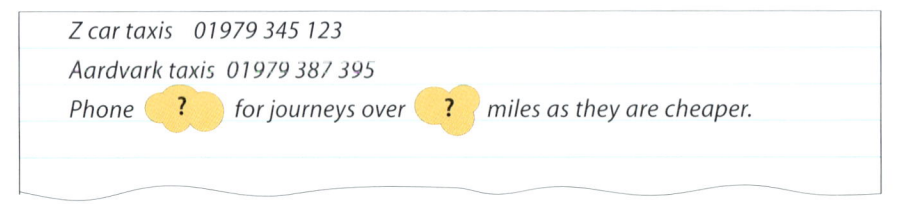

> Z car taxis   01979 345 123
> Aardvark taxis 01979 387 395
> Phone (?) for journeys over (?) miles as they are cheaper.

Test yourself

T1 Computer Helpline give help to computer users over the phone.
They use a formula to work out the charge.
The charge is based on how many minutes long the phone call is.

   charge in pounds = number of minutes × 3 + 5

So if a call is 2 minutes long, the charge is £2 × 3 + 5 = £11.

(a) Copy and complete this table.

Number of minutes	1	2	3	4	5	6
Charge in £		11				

(b) Draw a graph to show the values in your table.
Draw and label your axes like this.

Plot the points from your table and join them with a line.

Extend your line in both directions.

Go up to 28.
Stop at 7.

(c) Use the graph to say how much Computer Helpline charge for a $3\frac{1}{2}$ minute call.

(d) Alix makes a call. The charge is £21.50.
How long was her call?

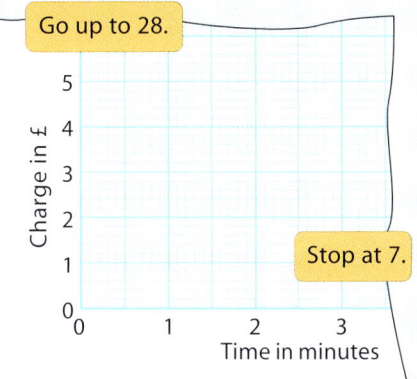

(e) Use your graph to find the missing numbers on these bills.

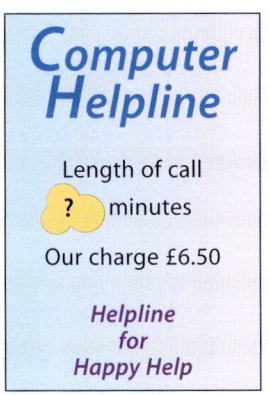

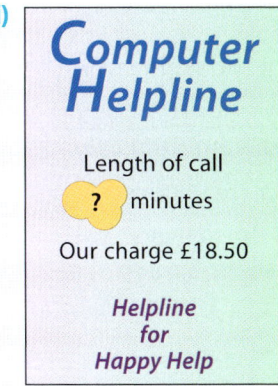

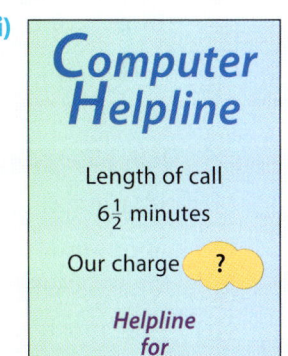

(i) **Computer Helpline**
Length of call
? minutes
Our charge £6.50
Helpline for Happy Help

(ii) **Computer Helpline**
Length of call
? minutes
Our charge £18.50
Helpline for Happy Help

(iii) **Computer Helpline**
Length of call
$6\frac{1}{2}$ minutes
Our charge ?
Helpline for Happy Help

34 Drawing and using graphs   157

Review 6

1. This shape is a prism made out of centimetre cubes. Find the volume of the prism in cm³.

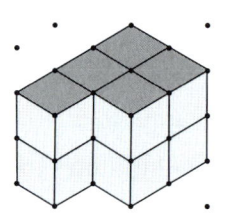

2. What is the value of each of the following expressions when $p = 8$?
   (a) $p - 1$
   (b) $3p - 1$
   (c) $3(p - 1)$
   (d) $\dfrac{p - 2}{3}$
   (e) $\dfrac{p}{2} - 3$

3. In a shop, all sale DVDs cost £5.99 each.
   (a) Kim has £20. How many of these DVDs can she buy?
   (b) Roughly, how much will 7 of these DVDs cost?

4. How many millilitres are in 2.8 litres?

5. Happy B Day supply birthday presents by post.
   They work out the delivery cost using the formula

   delivery cost in £ = weight of package in kg ÷ 2 + 3

   (a) What is the delivery cost of a package weighing 10 kg?
   (b) Copy and complete this table.

Weight in kg (w)	2	4	6	8	10	12
Delivery cost in £ (c)	4.00					

   (c) On graph paper, plot the points from your table, using axes like these.

   Join your points with a line and extend it.

   Use your graph to answer these questions.

   (d) About how much is the delivery cost for a package that weighs 3.4 kg?
   (e) How much is delivery for a 15 kg parcel?
   (f) Alice had to pay £7.25 delivery for a package. About how much did the package weigh?

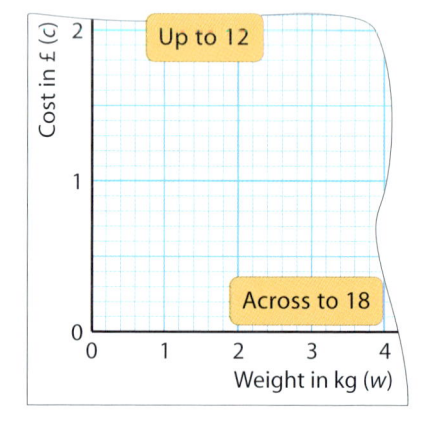

6. Which metric unit completes each statement?
   (a) A banana weighs 186 _____ .
   (b) A watering can holds 8 _____ .

7. Helen buys one chicken sandwich and four coffees for £8.90.
   A coffee costs £1.35.
   How much is a chicken sandwich?

35 Fractions, decimals and percentages

This work will help you
- find decimal and percentage equivalents of simple fractions
- find simple percentages of quantities without a calculator

You need sheets FT–23 and FT–24, and some counters.

A Fraction and percentage equivalents   level 4

0%   25%   50%   75%   100%

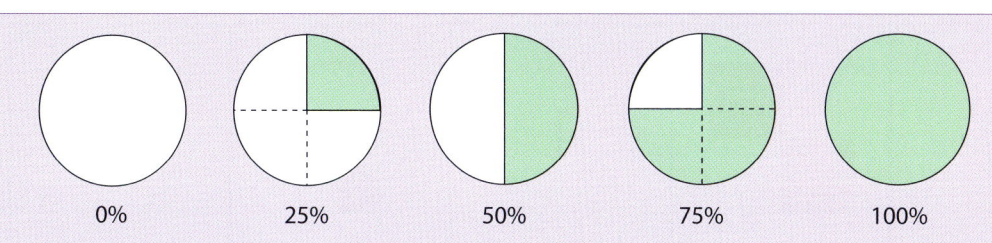

- What fraction of each of these circles is shaded?

- This circle is $\frac{1}{10}$ shaded.
  What **percentage** of the circle is shaded?

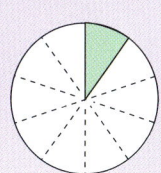

A1 Copy and complete these sentences.
  (a) $\frac{1}{2}$ is equivalent to ……%.
  (b) $\frac{1}{4}$ is equivalent to ……%.
  (c) … is equivalent to 10%.
  (d) … is equivalent to 75%.

A2 For each square, write down
  (i) the fraction shaded
  (ii) the percentage shaded
  (a)   (b) 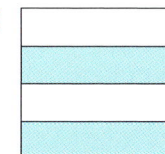   (c)

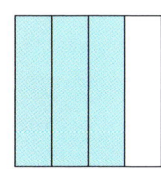

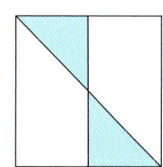

A3 Which is bigger, 10% or $\frac{1}{4}$?
  Explain your answer.

A4 Write each of these lists in order, starting with the smallest.
  (a) $\frac{3}{4}$, 100%, $\frac{1}{10}$, 50%
  (b) 75%, $\frac{1}{4}$, 10%, $\frac{1}{2}$

The FP game (fractions, percentages) for two players

You need an ordinary dice and counters (a different colour for each player).

- Take turns to roll the dice. This is what the faces mean.

  10%   25%   50%   75%   100%   Roll again

- Put a counter on any square with an equivalent expression or diagram on it. For example, if the dice says 10%, you could put a counter on $\frac{1}{10}$ or … (Only one counter is allowed on each square.)

- The winner is the first player with three of their counters in a line.

$\frac{3}{4}$		1		10%
50%	$\frac{2}{4}$			three quarters
25%			one quarter	$\frac{1}{10}$
	75%	one tenth	$\frac{1}{2}$	
$\frac{1}{4}$	one half	100%		

35 Fractions, decimals and percentages

B Fraction and decimal equivalents

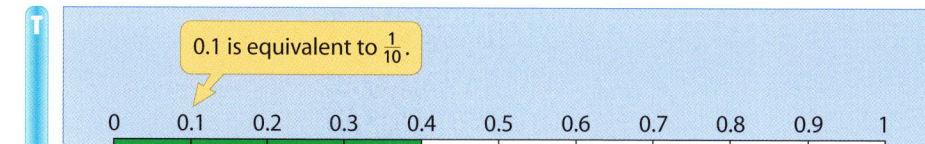

0.1 is equivalent to $\frac{1}{10}$.

0   0.1   0.2   0.3   0.4   0.5   0.6   0.7   0.8   0.9   1

- 0.4 of the bar is shaded. What is this as a fraction?
- What fraction is equivalent to 0.5?
- Complete the missing fractions on sheet FT–23.

B1 Write down a fraction equivalent to

(a) 0.2   (b) 0.9   (c) 0.3   (d) 0.7

B2 Find four matching pairs.

| 0.1 | $\frac{8}{10}$ | 0.5 | $\frac{6}{10}$ | $\frac{1}{10}$ | 0.6 | $\frac{5}{10}$ | 0.8 |

B3 Write these in order, starting with the smallest.

(a) $\frac{3}{10}$, 0.1, $\frac{9}{10}$, 0.8   (b) 0.6, $\frac{7}{10}$, 0.4, $\frac{2}{10}$

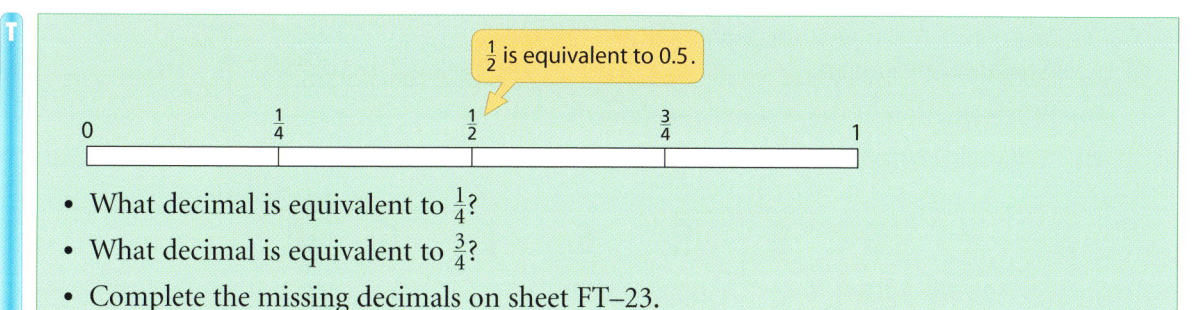

$\frac{1}{2}$ is equivalent to 0.5.

0    $\frac{1}{4}$ $\frac{1}{2}$ $\frac{3}{4}$ 1

- What decimal is equivalent to $\frac{1}{4}$?
- What decimal is equivalent to $\frac{3}{4}$?
- Complete the missing decimals on sheet FT–23.

Try to answer these without looking at the resource sheet.

B4 (a) Write down a decimal that is bigger than $\frac{1}{4}$ but smaller than $\frac{1}{2}$.
  (b) Write down a decimal that is bigger than $\frac{1}{2}$ but smaller than $\frac{3}{4}$.
  (c) Write down a decimal that is smaller than $\frac{1}{4}$.

B5 (a) A bag contains $\frac{1}{2}$ kg of sugar. Write this as a decimal of a kilogram.
  (b) A carton contains 0.25 litre of orange juice.
    Write this as a fraction of a litre.

B6 Write these in order, starting with the smallest.

(a) 0.75, $\frac{8}{10}$, $\frac{1}{2}$, 0.6   (b) $\frac{1}{2}$, 0.4, $\frac{1}{4}$, 0.3

C Fraction, decimal and percentage equivalents

C1 Fill in the fractions on sheet FT–24.

C2 Copy and complete this table.

Fraction		Decimal		Percentage
$\frac{1}{2}$	=		=	50%
	=	0.25	=	
$\frac{1}{10}$	=		=	
	=		=	20%

C3 (a) Sort these into three groups with equivalent values.

0.4   0.7 $\frac{4}{10}$ 75% $\frac{3}{4}$ 0.75   0.3   30%   40% $\frac{3}{10}$

(b) Which is the odd one out? Write down two values that are equivalent to it.

C4 (a) What fraction of this shape is shaded? 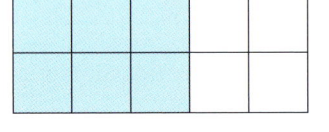
(b) Write this fraction as a decimal.
(c) What percentage of the shape is shaded?
(d) What percentage of the shape is not shaded?

C5 (a) Copy this rectangle onto squared paper. Shade seven squares. 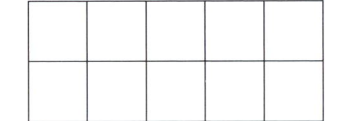
(b) What fraction of the rectangle is shaded?
(c) What percentage of the shape is shaded?

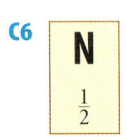

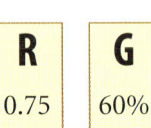

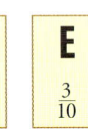

C6

N	Y	U	B	R	G	E	L	P
$\frac{1}{2}$	25%	0.8	$\frac{1}{10}$	0.75	60%	$\frac{3}{10}$	40%	0.2

Use this code to find a letter for each fraction, decimal or percentage below.
Rearrange each set of letters to spell a colour.

(a) 30%, $\frac{8}{10}$, 0.4, 0.1
(b) 75%, 0.6, $\frac{1}{4}$, 0.3
(c) 0.5, 30%, $\frac{6}{10}$, 75%, 0.3
(d) 20%, $\frac{8}{10}$, 0.4, 30%, $\frac{2}{10}$, $\frac{3}{4}$

C7 Which is bigger, 0.6 or 50%?
Explain your answer.

C8 Write these in order, starting with the smallest.

(a) 0.4, 20%, 60%, $\frac{1}{4}$
(b) $\frac{9}{10}$, 0.5, 80%, $\frac{3}{10}$

D Percentage of a quantity, mentally

Find 50% of 20.

> 50% is the same as $\frac{1}{2}$.
> $\frac{1}{2}$ of 20 = 20 ÷ 2 = 10
> so 50% of 20 is 10.

- How would you find 25% of 20?
- How would you find 75% of 20?

D1 Work out 50% of
(a) 8 (b) 12 (c) 40 (d) 100 (e) 200

D2 Work out 25% of
(a) 8 (b) 12 (c) 40 (d) 100 (e) 200

D3 Work out 75% of
(a) 8 (b) 12 (c) 40 (d) 100 (e) 200

Find 10% of 80.

> 10% is the same as $\frac{1}{10}$.
> $\frac{1}{10}$ of 80 = 80 ÷ 10 = 8
> so 10% of 80 is 8.

- How would you find 20% of 80?
- How would you find 30% of 80?

D4 Work out 10% of
(a) 50 (b) 30 (c) 200 (d) 250 (e) 140

D5 Work out 20% of
(a) 50 (b) 30 (c) 200 (d) 250 (e) 140

D6 Find four matching pairs that give the same answer.

| 10% of 40 | 20% of 50 | 50% of 6 | 20% of 20 |
| 10% of 30 | 25% of 60 | 25% of 40 | 50% of 30 |

D7 Work these out.

(a) 10% of 180 (b) 20% of 70 (c) 25% of 24 (d) 30% of 80 (e) 40% of 60

D8 Work these out.

(a) 10% of 40 (b) 5% of 40 (c) 15% of 40 (d) 30% of 40 (e) 35% of 40

D9 This box of cornflakes normally contains 500 g.
How many grams extra are in the box?

D10 In a raffle, 10% of tickets win prizes.
If 400 tickets are sold, how many prizes are won?

D11 A cinema has 300 seats.
One evening 15% of the seats are empty.
How many seats are empty?

D12 This pie chart shows how the pupils in year 7 travel to school.
There are 120 pupils in year 7.

(a) How many pupils walk to school?
(b) How many come by car?
(c) How many come by bus?
(d) How many cycle?

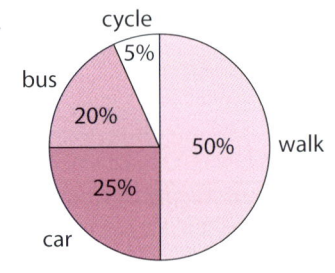

Test yourself

T1 (a) What fraction of this shape is shaded?
(b) What percentage of the shape is shaded?
(c) Copy the shape and shade in more squares so that $\frac{4}{5}$ of the shape is shaded.

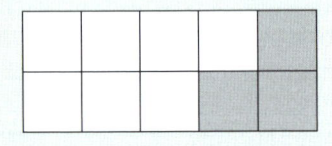

Edexcel

T2 (a) Write 50% as a fraction. (b) Write 0.6 as a fraction.
(c) Write $\frac{3}{4}$ as a decimal. (d) Write $\frac{7}{10}$ as a percentage.

T3 (a) Work out 10% of 60. (b) Work out 20% of 90.
(c) Work out 25% of 32. (d) Work out 5% of 80.

T4 In Year 7 there are 110 boys and 80 girls.

10% of the boys wear glasses.
15% of the girls wear glasses.

How many more girls than boys wear glasses?

OCR

36 Two-way tables

This work will help you
- read information from two-way tables
- make and use distance tables

You need sheet FT–25.

A Reading tables   level 3

Girls' names

This table shows the most popular names for baby girls in recent years in England and Wales.

Order	Year				
	2001	2002	2003	2004	2005
1	Chloe	Chloe	Emily	Emily	Jessica
2	Emily	Emily	Ellie	Ellie	Emily
3	Megan	Jessica	Chloe	Jessica	Sophie
4	Jessica	Ellie	Jessica	Sophie	Olivia
5	Sophie	Sophie	Sophie	Chloe	Chloe

Source: National Statistics

A1 What was the most popular name in 2004?

A2 What name was third most popular in 2002?

A3 (a) What is the highest position in the table reached by Sophie?
   (b) What year was this?

A4 In this table, which name appears for the first time in 2005?

Household jobs

A recent survey looked at which jobs adults liked doing around the home. The results are in the table.

Job	Male %	Female %
Cooking	58	69
Decorating	39	43
Gardening	47	50
Ironing	14	31
DIY repairs	50	21

Source: National Statistics

A5 What percentage of males liked decorating?

A6 What percentage of females liked ironing?

A7 Which of these statements are supported by this data? Write 'true' or 'false'.
   (a) Half of females liked gardening.
   (b) More than twice as many males as females liked ironing.
   (c) The least popular of these jobs for females was ironing.
   (d) Less than a quarter of females liked doing DIY repairs.

TV listings

Here are the listings for some of the main TV channels one afternoon.

BBC1	BBC2	ITV1	C4
1:40 Neighbours	1:30 Working Lunch	1:30 Loose Women	1:00 They Who Dare
2:05 Doctors	2:00 Narrow Escape	2:00 Ultimate Makeover	
2:35 Murder, She Wrote		3:00 CITV	3:00 Come Dine With Me
3:25 CBBC	3:30 Flog It!		3:30 Countdown
	4:30 Ready, Steady, Cook	4:30 Rising Damp	4:15 Deal or No Deal
5:35 Neighbours	5:15 The Weakest Link	5:00 Wycliffe	5:00 The Paul O'Grady Show
6:00 BBC News	6:00 Eggheads	6:00 Central News	6:00 The Simpsons

A8 Which programme starts at 3:30 on Channel 4 (C4)?

A9 How long are each of these programmes?

(a) Loose Women (ITV1) (b) Deal or No Deal (C4) (c) Flog It! (BBC2)
(d) Neighbours (BBC1) (e) They Who Dare (C4) (f) CBBC (BBC1)

A10 For each channel, write down the programme that is on at 4:00.

B Distance tables   level 4

This map shows the main towns in the Republic of Ireland.

The distances shown are the shortest distances by road between the places.

- What is the shortest distance between Longford and Limerick?
- What is the shortest distance between Dublin and Limerick?

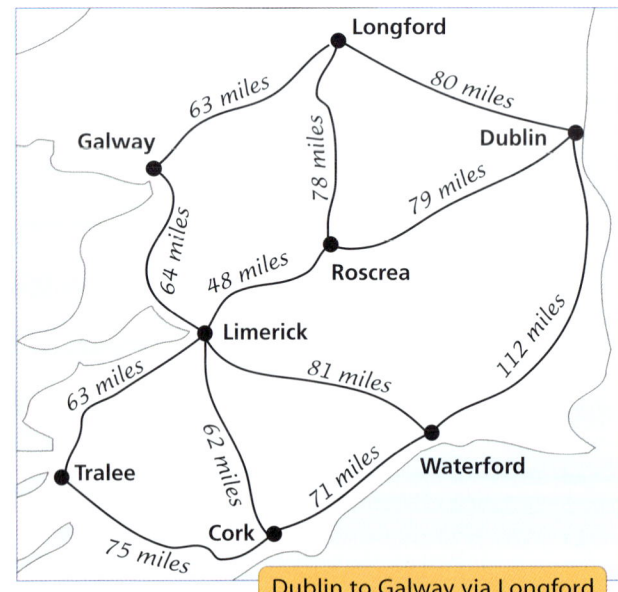

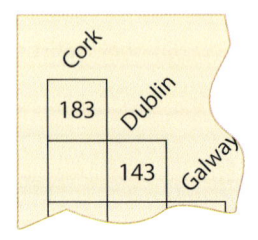

Longford to Dublin: 80 miles
Galway to Longford: 63 miles
Galway to Roscrea: 78 miles
Dublin to Roscrea: 79 miles
Galway to Limerick: 64 miles
Limerick to Roscrea: 48 miles
Roscrea to Waterford: 112 miles
Limerick to Waterford: 81 miles
Limerick to Tralee: 63 miles
Tralee to Cork: 75 miles
Limerick to Cork: 62 miles
Cork to Waterford: 71 miles

Dublin to Galway via Longford is 80 + 63 = 143 miles.

This is part of a table showing the **shortest** distance between places.

	Cork	Dublin	Galway
	183		
		143	

B1 Use sheet FT–25 and complete the distance table.

B2 (a) Maria drives from Limerick to Cork.
How far does she drive?

(b) She then drives from Limerick to Waterford and then back to Cork.
How far does she drive?

(c) How far has she driven altogether?

This table shows the rough distances in miles between some of the main UK airports.

B3 (a) How far is it between East Midlands and Heathrow?

(b) Which two airports in this table are the furthest apart?

B4 Roger flies from Gatwick to Glasgow and back.
On the way there he flies direct.
On the way back he flies via Manchester.

(a) How far did he fly going to Glasgow?

(b) How far did he fly coming back in total?

(c) How much further was the journey coming back?

	Aberdeen	East Midlands	Gatwick	Glasgow	Heathrow	Manchester
East Midlands	295					
Gatwick	585	155				
Glasgow	155	295	445			
Heathrow	555	120	40	405		
Manchester	360	80	235	225	200	

Test yourself

T1 The distances, in miles, between seven cities in Britain are given in the chart below.

Cardiff						
302	Carlisle					
401	172	Edinburgh				
155	313	413	London			
192	119	219	204	Manchester		
202	164	247	169	37	Sheffield	
141	343	442	80	233	207	Southampton

(a) Kim drives from Cardiff to Carlisle.
What is the distance from Cardiff to Carlisle?

(b) Kim then drives from Carlisle to Sheffield.
What is the distance from Carlisle to Sheffield?

(c) Kim then returns to Cardiff directly from Sheffield.
How many miles does Kim travel in total?

AQA

37 Scale drawings

This work will help you make scale drawings and interpret them.

You need an angle measurer.

A Simple scales

B This drawing on centimetre squared paper is a **scale drawing** of the blue shape.

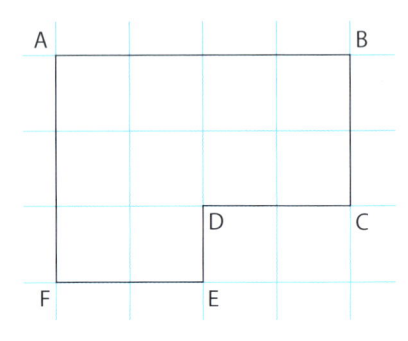

A1 (a) Copy this table.

Measure the sides of the blue shape and its scale drawing. Record their lengths in your table. Two measurements have been done for you.

Side	Length on blue shape	Length on scale drawing
AB	8 cm	4 cm
BC		
CD		
DE		
EF		
FA		

(b) What does 1 cm on the scale drawing represent on the blue shape?

(c) Measure the length of the diagonal BF on the scale drawing.

(d) How long should diagonal BF be on the blue shape? Measure to see whether you were right.

A2 (a) On centimetre squared paper, make a scale drawing of this orange square.
Use a scale where 1 cm on the scale drawing represents 2 cm on the orange square.

(b) Measure the diagonal of the orange square, to the nearest 0.1 cm.

(c) How long should the diagonal be on the scale drawing? Measure to see whether you were right.

168   37 Scale drawings

A3 This is a scale drawing of a warning sign.
1 cm on the scale drawing represents 2 cm on the real sign.

(a) What is the height of the real sign?

(b) What is the width of the real sign?

(c) Measure the diagonal of the scale drawing, to the nearest 0.1 cm.

(d) What is the length of the diagonal of the real sign?

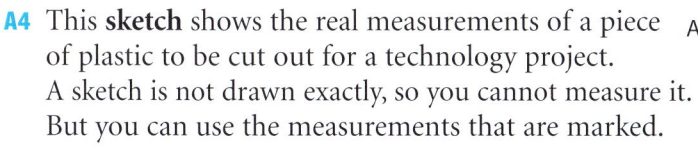

A4 This **sketch** shows the real measurements of a piece of plastic to be cut out for a technology project. A sketch is not drawn exactly, so you cannot measure it. But you can use the measurements that are marked.

(a) On centimetre squared paper, make a scale drawing of the piece of plastic, using a scale where 1 cm represents 2 cm.

(b) On the piece of plastic, will the length AD be more or less than 14 cm?

(c) Measure AD on your scale drawing to the nearest 0.1 cm.

(d) How long will AD be on the piece of plastic?

A5 This is a sketch of a garden seen from above.

(a) On centimetre squared paper, make a scale drawing of the garden, using a scale where 1 cm represents 2 **metres**.

(b) Measure the length XY on your scale drawing.

(c) How long will XY be on the actual garden? Remember to give your answer in metres.

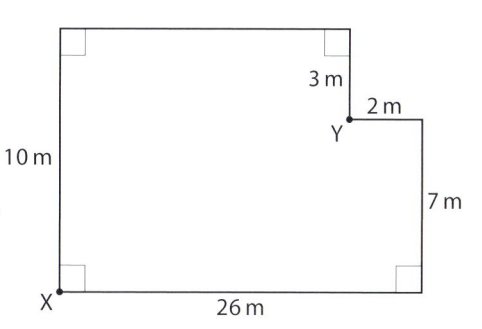

A6 The triangle on centimetre squared paper is a scale drawing of the blue triangle.

(a) What is the height of the triangle on the scale drawing?

(b) What is the height of the blue triangle?

(c) Write the scale of the scale drawing as '1 cm represents ___ cm.'

(d) What is the length of the base of the triangle on the scale drawing?

(e) How long should the base of the blue triangle be? Measure to see whether you were right.

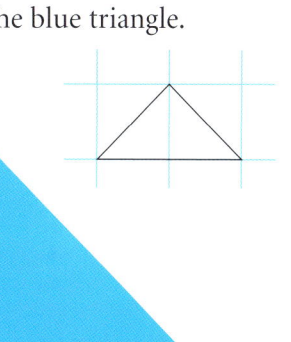

A7 This is a sketch of a shape to be cut from cloth fabric.

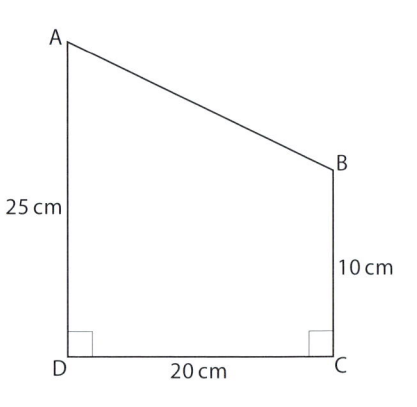

(a) On centimetre squared paper, make a scale drawing of the fabric shape, using a scale where 1 cm represents 5 cm.

(b) Measure the length AB on your scale drawing.

(c) How long will AB be on the real fabric shape?

(d) Measure the acute angle at corner A on your scale drawing.

A8 This is a sketch plan of a field.

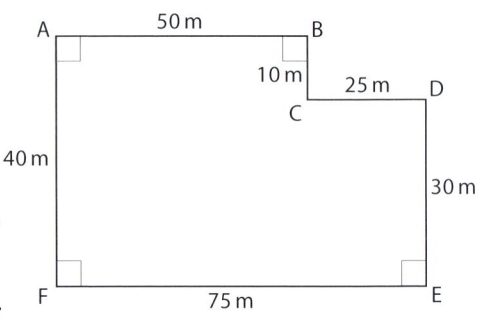

(a) On centimetre squared paper, make a scale drawing of the field, using a scale where 1 cm represents 5 **metres**.

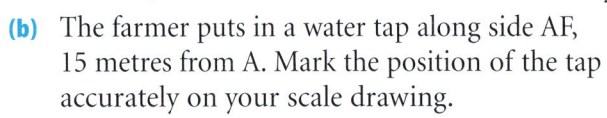

(b) The farmer puts in a water tap along side AF, 15 metres from A. Mark the position of the tap accurately on your scale drawing.

(c) A footpath goes in a straight line from A to E. Measure AE on your scale drawing.

(d) How long is the footpath on the real field?

A9 This is a sketch of a shape to be cut from plywood.

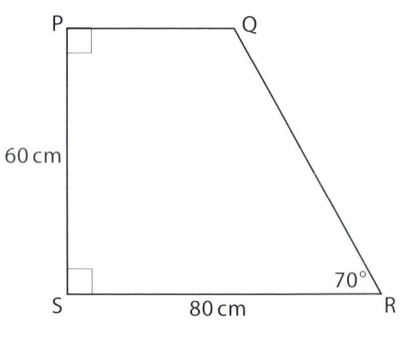

(a) On centimetre squared paper, make a scale drawing of it, using a scale where 1 cm represents **10** cm.

(b) Measure the length PQ on your scale drawing, to the nearest 0.1 cm.

(c) How long will PQ be on the real plywood shape?

(d) Measure the length QR on your scale drawing, to the nearest 0.1 cm.

(e) How long will QR be on the real plywood shape?

A10 This is a map of an island, drawn to a scale where 1 cm represents 10 **kilometres**.

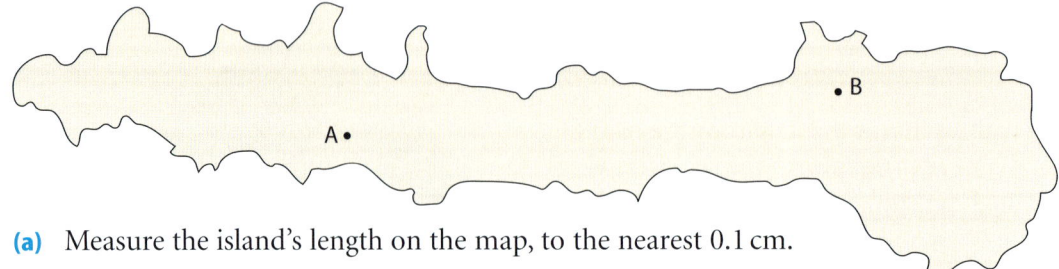

(a) Measure the island's length on the map, to the nearest 0.1 cm.

(b) What is the length of the real island?

(c) The points A and B are two radio masts. What is the real distance between A and B?

B Harder scales

B1 This is a scale plan of a room.

2 centimetres on the plan stands for 1 metre in the real room.
We say the scale is '2 centimetres to 1 metre'.

On the plan, the width of the room is 6 cm.
So the width of the real room is 3 m.

(a) Measure the length of the room on the plan.

(b) What is the length of the real room?

(c) Measure the width of the window on the plan.

(d) What is the width of the real window?

(e) Measure the width of the archway on the plan.

(f) What is the width of the real archway?

(g) A rectangular table 1 metre wide by 1.5 metres long is brought into the room.
If it was shown on this scale drawing, what size would it be drawn?

B2 This is a scale drawing of a sideboard.

5 centimetres on the scale drawing stands for 1 metre on the real sideboard.
We say the scale is '5 centimetres to 1 metre'.

(a) Measure the width of the sideboard on the drawing.

(b) What is the width of the real sideboard?

(c) What does 1 cm on the drawing stand for on the real sideboard?

(d) Measure the length of the sideboard's legs on the drawing.

(e) How long are the legs of the real sideboard?

(f) What is the distance from the floor to the top of the real sideboard?

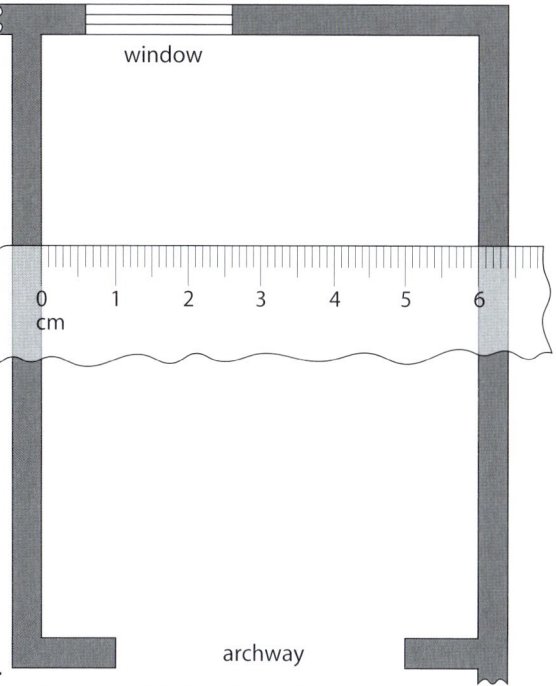

Test yourself

T1 This is a scale drawing of a shape to be cut from cloth fabric.
1 cm on the scale drawing represents 5 cm on the actual shape.

(a) What is the length of diagonal AC on the real shape?

(b) Measure the length of side BC on the scale drawing.

(c) What will be the length of side BC on the real shape?

(d) A circle with diameter 15 cm is to be screen-printed on the fabric. What would the diameter of the circle be on the scale drawing?

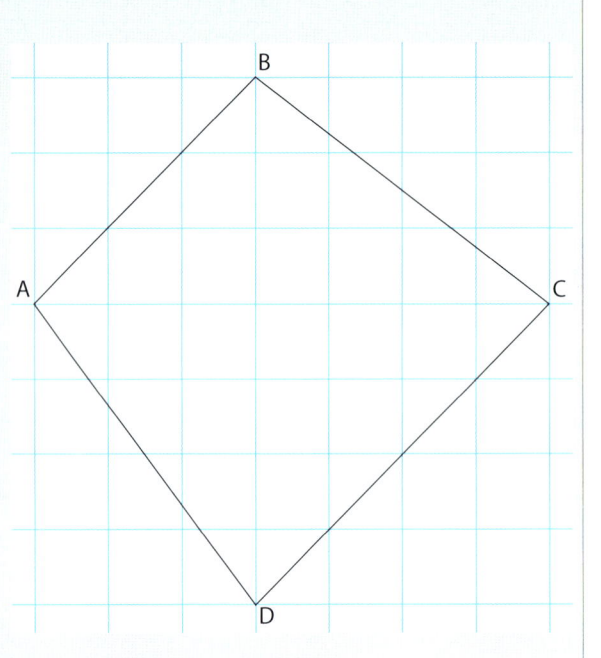

T2 This is a scale drawing of a triangular flag.
1 cm on the scale drawing represents 10 cm on the actual flag.

(a) What is the length of the side PR on the real flag?

(b) Measure the length of side PQ on the scale drawing of the flag, to the nearest 0.1 cm.

(c) What will be the length of side PQ on the real flag?

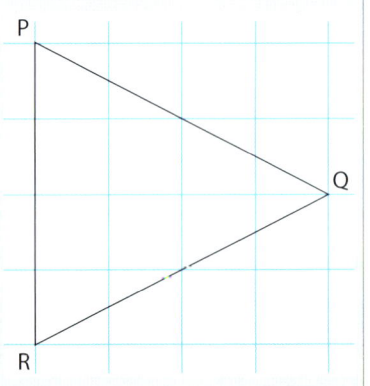

T3 The diagram shows the side view of a building.

(a) Using a scale of 2 cm to represent 1 m, make an accurate drawing of the side view of the building.

(b) Use your scale drawing to find the **real** length r.

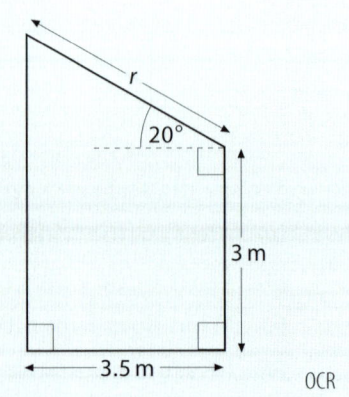

Not to scale

OCR

172 37 Scale drawings

38 Using percentages

This work is about using different charts to present data.

It will help you
- draw pie charts using percentages
- interpret and draw composite bar charts

You need a pie chart scale.

A Percentage bars

Here is some information given on a tub of ice cream.

Nutritional information
Energy   167 kcal per 100 g
Water   64%
Sugar   23%
Starch   2%
Protein   4%
Fat   7%

This information can be shown on a bar.

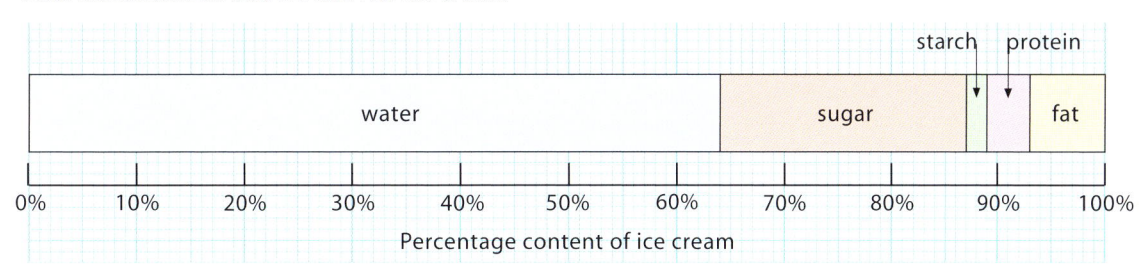

Percentage content of ice cream

A1 This diagram shows the contents of a typical cheddar cheese.

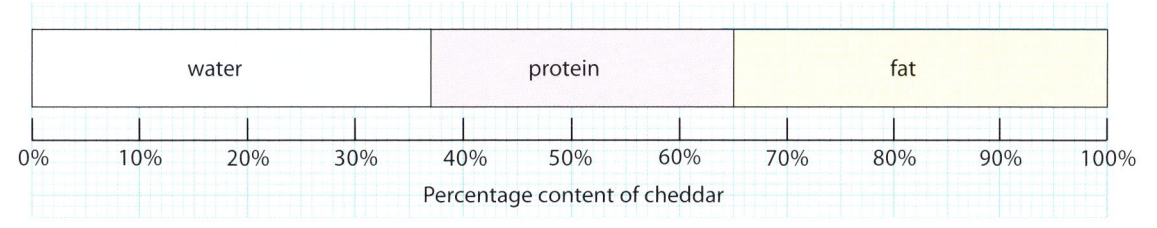

Percentage content of cheddar

(a) What percentage of cheddar cheese is water?
(b) What percentage of cheddar cheese is not water?
(c) What percentage of cheddar cheese is protein?
(d) What percentage of cheddar cheese is fat?

A2 These percentages show the contents of cottage cheese. Use this information to draw a chart on graph paper showing the contents of cottage cheese.

Contents
Water   80%
Protein   15%
Fat   5%

B Interpreting pie charts

Information that uses percentages can be shown in a round diagram called a **pie chart**.

To read from or draw a pie chart you will need a pie chart scale.

Each time you want to measure a new section of a pie chart, you must place the 0% on the start of a section and measure clockwise.

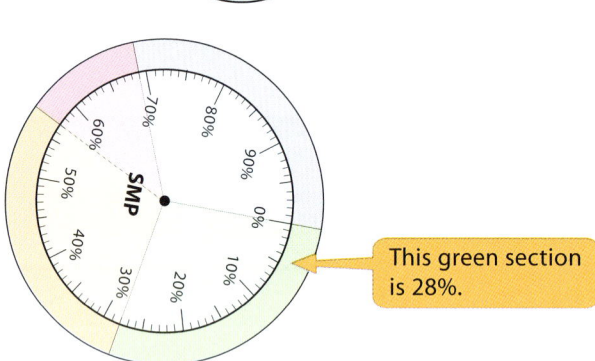

This green section is 28%.

Use a pie chart scale to answer these questions.

B1 This pie chart shows the contents of an egg.
  (a) What percentage of water does it contain?
  (b) What percentage of protein does it contain?
  (c) What percentage of fat does it contain?
  (d) Write 'true' or 'false' for each of these statements.

  A An egg is almost 90% fat free.
  B Around three quarters of an egg is water.
  C There is more fat than protein in an egg.

  (e) An egg weighs about 60 g. Roughly how many grams of this is water?
  (f) Roughly how much of a 60 g egg is fat?

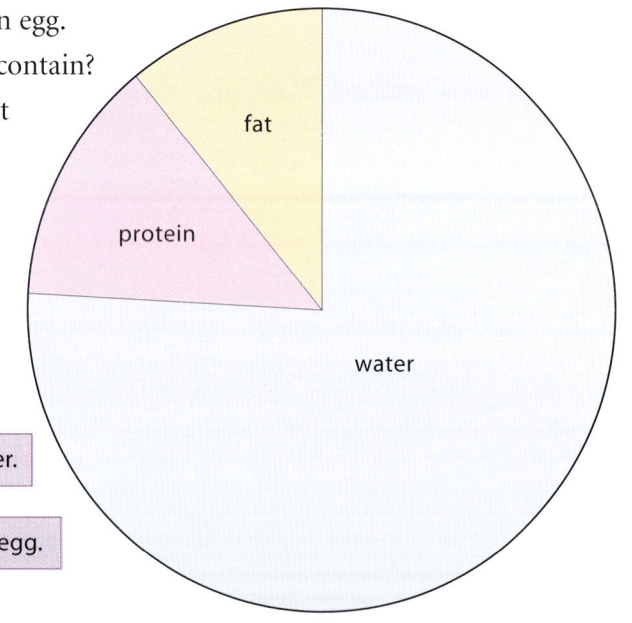

B2 Gary keeps a record of what he spends his money on over one month.
This pie chart shows what he spent money on.

(a) What did Gary spend the most money on?
(b) What percentage of his money did he spend on food and drink?
(c) What percentage of his money did he spend on going out and CDs?
(d) If Gary spent £40 that month, roughly how much did he spend on going out and CDs?

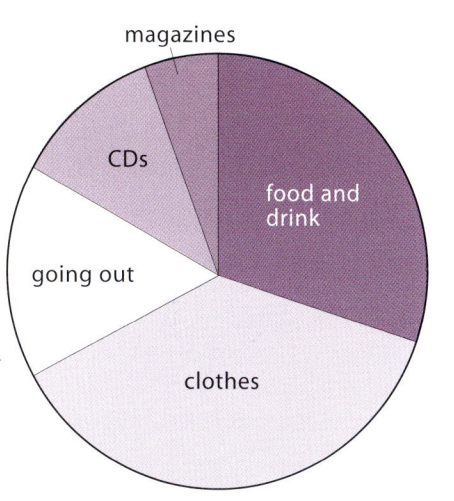

B3 This pie chart shows how the UK government spent taxpayers' money in 2004/5.

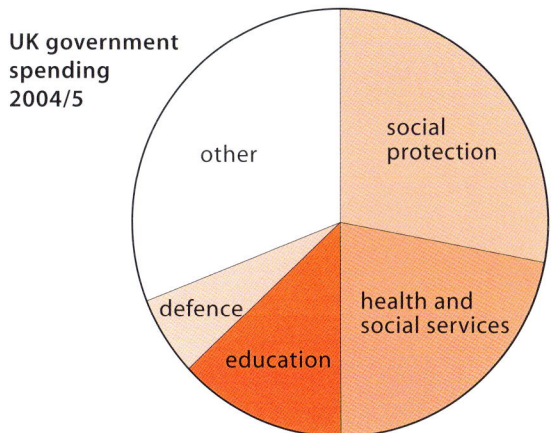

(a) What percentage did the government spend on social protection?
(b) What percentage did the government spend on education?
(c) Are these statements true or false?
   A In 2004/5 the government used just over a quarter of all its spending for social protection.
   B In 2004/5 the government used half of all its spending on health and social services.
   C In 2004/5 the government spent about twice as much on education as defence.

B4 This pie chart shows what a group of people in the UK said they spent their earnings on.

(a) What did these people say they spent their money on most of all?
What percentage is this?
(b) Did these people spend more on food and drink or on transport?
(c) What percentage did these people spend on leisure?

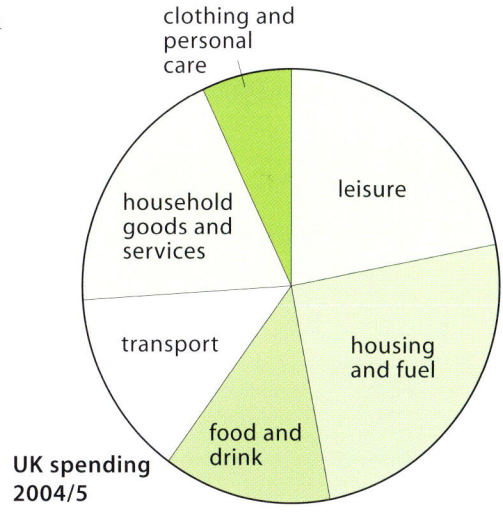

38 Using percentages 175

C Drawing pie charts

The contents of a pizza are

Water 48%   Protein 9%   Fat 16%   Carbohydrate 27%

To draw a pie chart of this information…

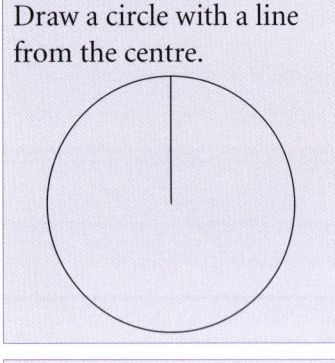

Draw a circle with a line from the centre.	Put the scale on the centre with 0% on the line. Mark at 48%.	Draw another line. Label.

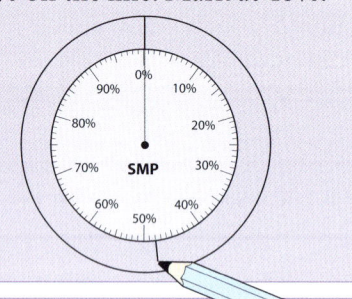

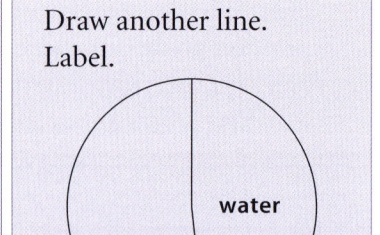

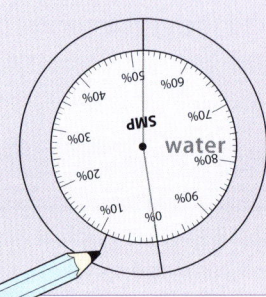

For the protein section, put the 0% on the line you have just drawn. Mark at 9% and draw the line.
Repeat for each section.

C1 Copy and complete the pie chart showing the contents of a pizza.
Use a circle with radius 4 cm.
Label each section of your chart.

C2 This information shows the crimes recorded by the police in England and Wales in 2004/5.

| Theft 36% | Criminal damage 21% | Violence 19% |
| Burglary 12% | Fraud 5% | Other 7% |

Use this information to draw a pie chart with radius 4 cm showing the crimes recorded in 2004/5.

C3 This information shows where the UK sold its goods abroad (its exports) in 2003.

	European Union	Other European	North America	Others
Exports	58%	11%	12%	19%

Draw a pie chart with radius 4 cm to show where the UK exports its goods.

D Comparing

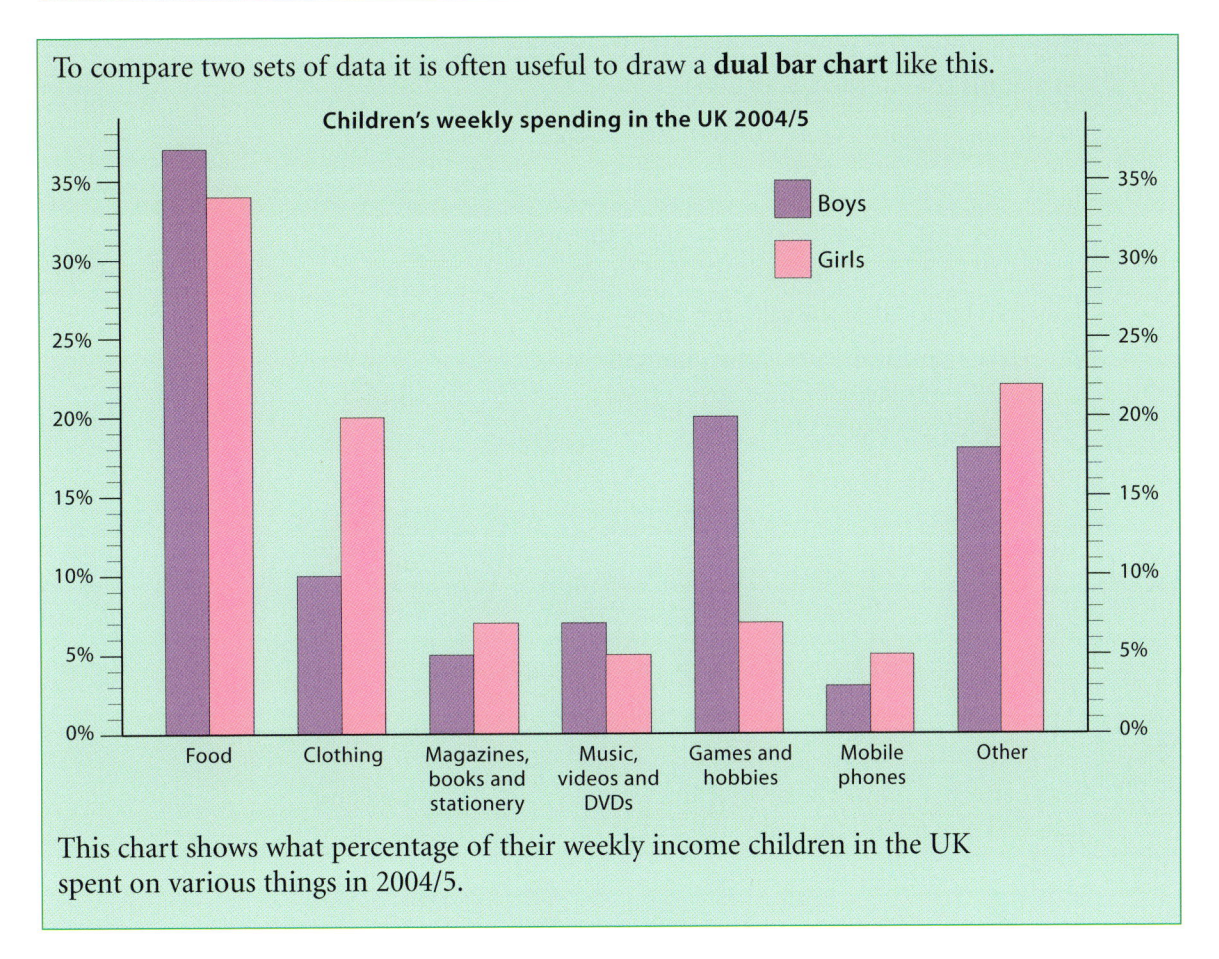

To compare two sets of data it is often useful to draw a **dual bar chart** like this.

Children's weekly spending in the UK 2004/5

[Bar chart showing Boys and Girls percentages across categories:
- Food: Boys ~37%, Girls ~34%
- Clothing: Boys ~10%, Girls ~20%
- Magazines, books and stationery: Boys ~5%, Girls ~7%
- Music, videos and DVDs: Boys ~7%, Girls ~5%
- Games and hobbies: Boys ~20%, Girls ~7%
- Mobile phones: Boys ~3%, Girls ~5%
- Other: Boys ~18%, Girls ~22%]

This chart shows what percentage of their weekly income children in the UK spent on various things in 2004/5.

D1 **(a)** What percentage of their income did boys spend on games and hobbies?

(b) What percentage of their income did girls spend on food?

D2 What items did girls spend a higher percentage on than boys?

D3 These figures show the contents of two types of cheese.

	Water	Protein	Fat	Other
Camembert	48%	23%	23%	6%
Parmesan	28%	35%	30%	7%

Use this information to draw a dual bar chart comparing the contents of the two cheeses.

Describe the differences in the contents of the two cheeses.

Test yourself

T1 This table gives information about the types of music on albums sold in the UK in 2000.

Music type	Rock	Pop	Urban	MOR	Dance	Other
UK sales	26%	32%	12%	5%	13%	12%

(a) Draw and label a pie chart to show this information.

This pie chart shows the sales for 2005.

(b) Make two comments about changes in the popularity of the music types between 2000 and 2005.

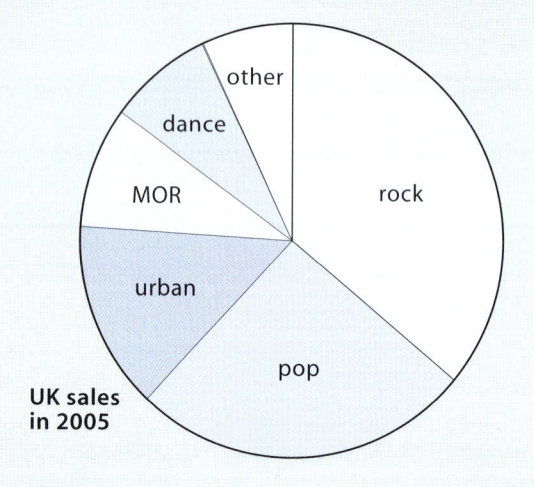

UK sales in 2005

T2 This bar chart shows the days of the week a group of women and a group of men said they did shopping.

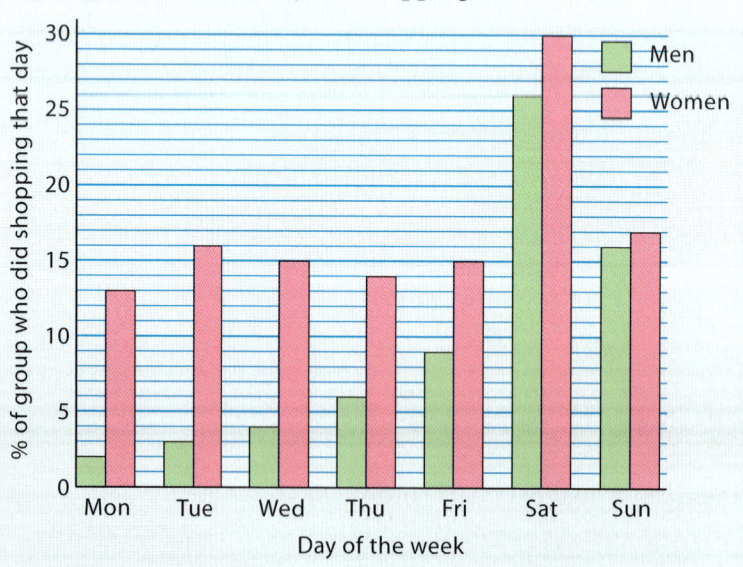

(a) Which is the most popular shopping day for the men?

(b) Give two ways in which the women's shopping is different from the men's.

178 38 Using percentages

39 Conversion graphs

This work will help you use and draw a conversion graph.

You need sheets FT–26 and FT–27.

A Using a conversion graph

This graph can be used to convert between miles and kilometres.

For example, the dotted line shows that 12 miles is about 19 km.

A1 Use the graph to convert

(a) 20 miles to km

(b) 27 miles to km

(c) 8 miles to km

(d) 17.5 miles to km

A2 Find 35 kilometres on the vertical scale of the graph.
Use the graph to convert 35 km to miles.

A3 Use the graph to convert

(a) 40 km to miles

(b) 24 km to miles

(c) 10 km to miles

A4 You can't use the graph to convert 50 miles to km, because the 'miles' scale doesn't go up to 50.

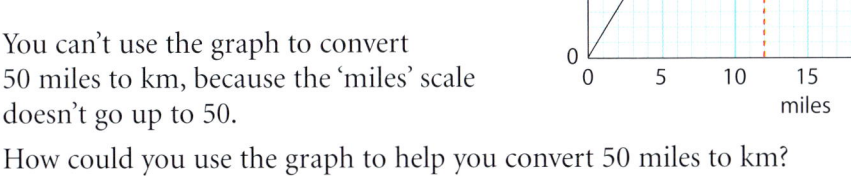

How could you use the graph to help you convert 50 miles to km?

A5 Use the graph to help you convert

(a) 100 miles to km

(b) 75 miles to km

(c) 60 km to miles

(d) 100 km to miles

A6 The distance from London to Cambridge is 55 miles.
How much is this in kilometres?

A7 This graph can be used to convert between pints and litres.

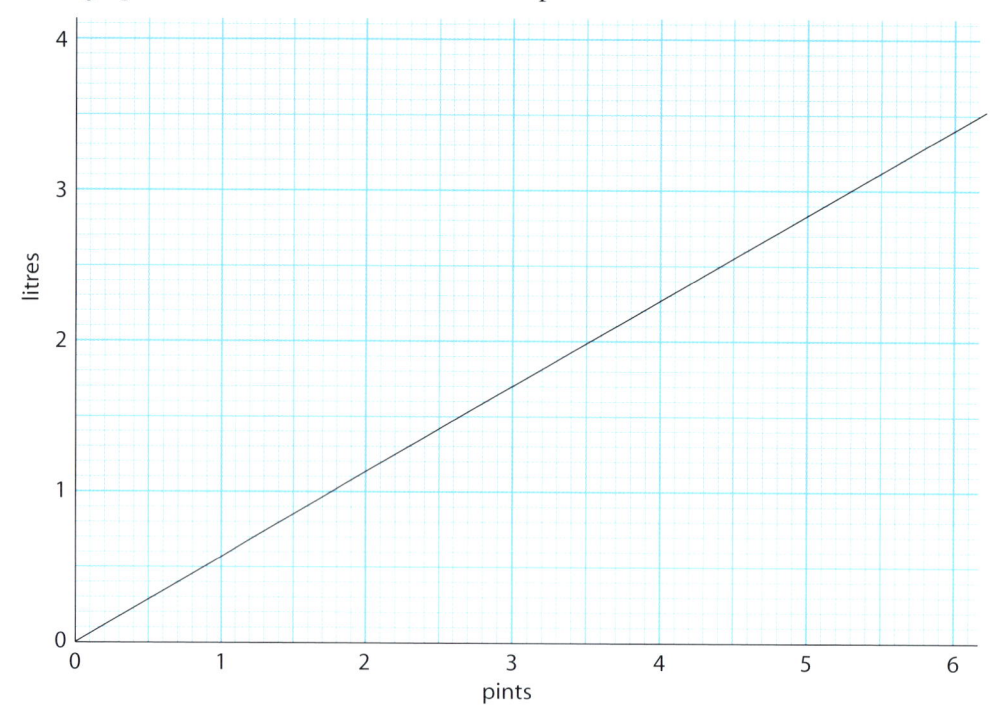

(a) Use the graph to convert
 (i) 4.5 pints to litres
 (ii) 2.8 pints to litres
 (iii) 1.7 pints to litres

(b) Use the graph to convert
 (i) 3 litres to pints
 (ii) 1.5 litres to pints
 (iii) 0.9 litre to pints

(c) Use the graph to help you convert 15 pints to litres.

A8 (a) Use this conversion graph to convert
 (i) 4 pounds to kilograms
 (ii) 5.5 pounds to kilograms
 (iii) 2.4 kg to pounds
 (iv) 0.9 kg to pounds

(b) Use the graph to help you convert
 (i) 12 pounds to kilograms
 (ii) 10 kilograms to pounds

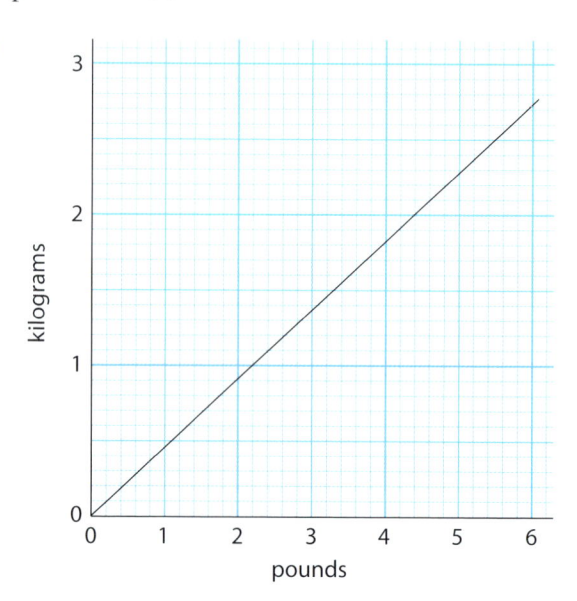

B Drawing a conversion graph

You need sheet FT–26.

B1 Ships' speeds can be measured in knots.

This table shows speeds in knots and in kilometres per hour (km/h).

Speed in knots	0	5	10	15	20	25
Speed in km/h	0	9	18	27	36	45

(a) Plot the points from the table on grid A on the sheet. Draw the straight line through the points.

(b) Use your graph to convert

 (i) 17 knots to kilometres per hour

 (ii) 39 kilometres per hour to knots

B2 This table shows areas in square miles and km².

square miles	0	10	20	30	40	50
km²	0	26	52	78	104	130

(a) On grid B draw a conversion graph.

(b) Use the graph to convert

 (i) 37 square miles to km² (ii) 85 km² to square miles

B3 This table shows volumes in cubic feet and m³.

cubic feet	0	50	100	150	200	250
m³	0	1.4	2.8	4.2	5.6	7.0

(a) On grid C draw a conversion graph.

(b) Use the graph to convert

 (i) 220 cubic feet to m³ (ii) 5.2 m³ to cubic feet

B4 An airport shop shows its prices in pounds sterling (£) and euros (€). A compact camera costs £25 or €40.

Compact camera
£25   €40

(a) On grid D mark the point (25, 40). The conversion graph must start at (0, 0) because £0 = €0. Draw the straight line from (0, 0) through (25, 40).

(b) Use your conversion graph to find the missing prices on these tickets.

(i) Digital watch
£17.50   €……

(ii) Make-up bag
£……   €35

(iii) Perfume
£……   €19

B5 You need sheet FT–27.

The following table shows how °Celsius can be changed into °Fahrenheit.

°C	-5	0	5	10	15	20	25	30	35
°F	23	32	41	50	59	68	77	86	95

(a) Use the above information to draw, on grid A, a conversion graph for changing °C to °F.

Use your graph to

(b) change 80 °F to °C   (c) change ⁻10 °C to °F

WJEC

Test yourself

T1 This conversion graph is for dollars ($) to pounds (£).

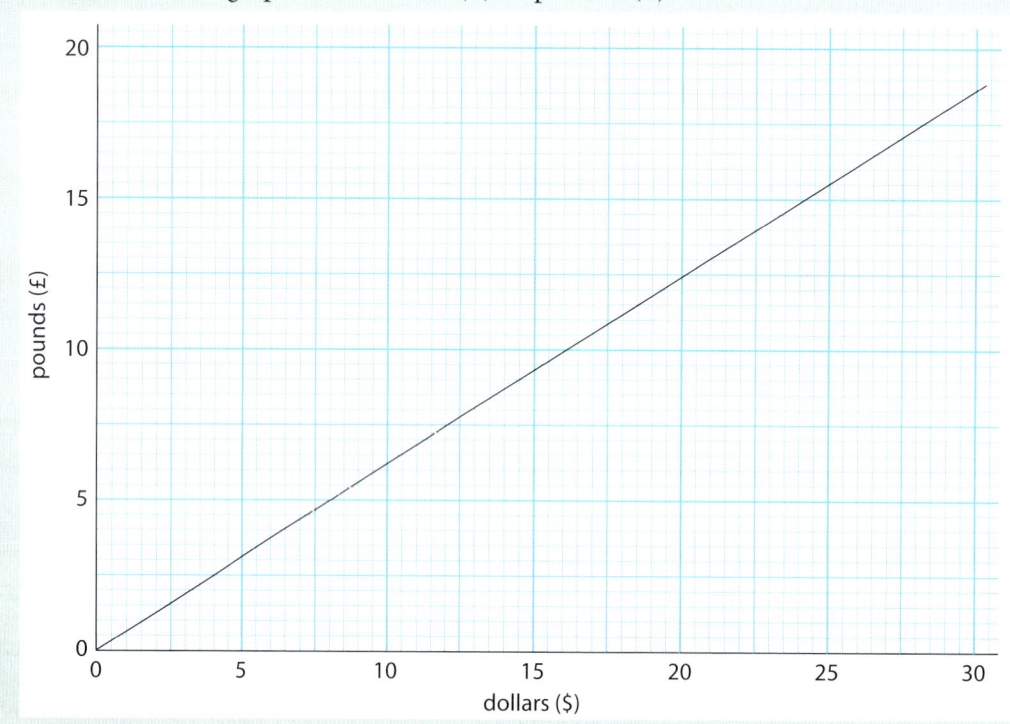

Use the graph to

(a) change £5 into dollars ($)   (b) change $17 into pounds (£)

OCR

T2 You need sheet FT–27.

5 litres is equal to 1.1 gallons and 10 litres is equal to 2.2 gallons.

(a) Plot this information on grid B in order to produce a conversion graph.

(b) Use your graph to convert   (i) 8 litres to gallons   (ii) 3 gallons to litres

(c) Estimate how many litres there are in half a gallon.

182 39 Conversion graphs

Review 7

You need a pie chart scale for question 3.

1 Write

(a) $\frac{3}{10}$ as a decimal   (b) 0.75 as a fraction   (c) 40% as a fraction

2 The distance in miles between five Scottish towns and cities is given in the chart below.

(a) What is the distance from Edinburgh to Dundee?

(b) Gerry drives from Edinburgh to Glasgow and then goes on from Glasgow to Oban. How many miles does he travel in total?

Glasgow				
44	Edinburgh			
84	124	Stranraer		
83	56	167	Dundee	
92	123	148	117	Oban

This graph can be used to convert between miles and kilometres.

(c) Use the graph to estimate the distance to the nearest 10 kilometres between

(i) Stranraer and Oban

(ii) Glasgow and Dundee

(d) The distance between Inverness and Glasgow is 267 kilometres. Use the graph to find an estimate of this distance in miles.

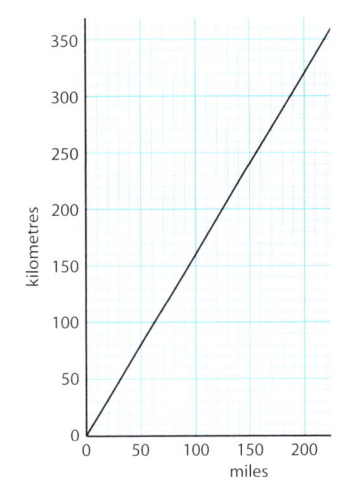

3 This data is about the total number of people killed in road accidents in the UK in 2000.

	Pedestrians	Cyclists	Motorcyclists	All other road users
Number killed	25%	4%	18%	53%

(a) What fraction of the people killed were pedestrians?

(b) Draw a pie chart to show this data.

4 This is a sketch of a garden seen from above.

(a) On cm squared paper, make a scale drawing of the garden using a scale where 1 cm represents 2 metres.

(b) Measure the length PQ on your scale drawing.

(c) How long will PQ be on the actual garden?

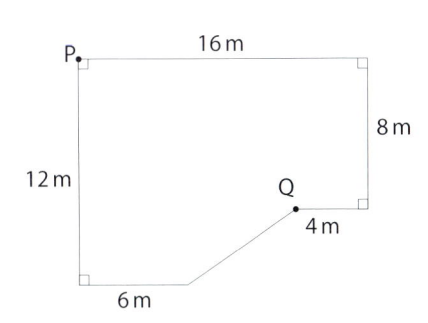

Answers

1 Odds, evens, multiples, factors

A Odd and even numbers (p 8)

A1 (a) 6, 14, 20   (b) 7, 11, 21

A2 (a) Bear   (b) Bunny   (c) Bear
   (d) Bear   (e) Bunny

A3 The letters are E, I, M, L and they rearrange to give LIME.

A4 An odd number

B Divisibility (p 9)

B1 (a) 25, 30, 50, 55, 80
   (b) 16, 30, 50, 52, 80
   (c) (i) 30, 50, 80   (ii) They are divisible by 10.

B2 (a) Balloon   (b) £1   (c) Balloon
   (d) £1   (e) Lollipop

B3 (a) (i) 7
       (ii) There is no remainder so 21 is divisible by 3.
   (b) 6, 12, 18, 24, 30
   (c) 4, 8, 12, 24

B4 (a) (i) O, D, G   (ii) G, E, R, I, T
       (iii) O, A, G, T   (iv) B, A, E, R
   (b) (i) DOG   (ii) TIGER
       (iii) GOAT   (iv) BEAR

B5 (a) 45, 180   (b) 45, 47, 89, 91, 93
   (c) 45   (d) 184 and 46

C Multiples (p 10)

C1 5, 10, 15, 20, **25, 30**, 35, **40, 45**, 50

C2 4, 8, 12, 16, 20, 24, 28

C3 50, 20, 130, 240

C4 (a) 12, 15, 30, 36, 39   (b) 15, 30, 50, 65
   (c) 12, 30, 36

C5 (a) (i) O, N, I, B, R   (ii) O, W, L
       (iii) O, C, R, W   (iv) S, N, A, W
   (b) (i) ROBIN   (ii) OWL
       (iii) CROW   (iv) SWAN

C6 (a) 14   (b) 9   (c) 24   (d) 45   (e) 35

D Factors (p 11)

D1 3

D2 2, 1, 3, 4, 6

D3 1, 2, 3, 6

D4 2, 5, 1, 10, 4

D5 (a) 3   (b) 7   (c) 1

Test yourself (p 11)

T1 (a) **3, 57, 65 and 71** are all the odd numbers.
   (b) **50 and 65** can be divided by five exactly.
   (c) **50** can be divided by ten exactly.
   (d) **3 and 57** add up to 60.

T2 (a) 2, 6, 30, 40   (b) 30, 40
   (c) 6, 9, 15, 30   (d) 1, 2, 6, 9

T3 (a) 33   (b) 50

2 Mental methods 1

A Place value (p 12)

A1 (a) 3000 (b) 50
A2 (a) 400 000 (b) 90 000
A3 (a) 3618 (b) 52 763 (c) 42 371
(d) 148 510 (e) 90 284 (f) 473 593
(g) 95 342 (h) 274 540
A4 2000
A5 (a) 1911 (b) 3711 (c) 21 711
A6 (a) (i) 8743
(ii) Eight thousand seven hundred and forty-three
(b) (i) 3478
(ii) Three thousand four hundred and seventy-eight
A7 (a) Graham Gooch
(b) Mike Atherton
(c) 8900, 8463, 8231, 8114, 7728
A8 (a) 1005, 1050, 1500, 5010, 5100
(b) 67 275, 67 342, 67 531, 68 200, 68 752
(c) 20 840, 24 030, 38 040, 40 250, 41 000

B Rounding (p 13)

B1 (a) 80 (b) 30 (c) 440 (d) 910 (e) 3290
B2 (a) 500 (b) 400 (c) 2600 (d) 5800 (e) 32 400
B3 (a) 7000 (b) 8000 (c) 3000 (d) 18 000 (e) 8000
B4 (a) 4100 (b) 4130 (c) 4000
B5 (a) 12 000 (b) 10 000 (c) 11 700 (d) 11 680
B6 (a) 46 000 (b) 68 100 (c) 650 000 (d) 8 000 000
B7 (a)

Ocean	Greatest depth
Pacific Ocean	11 000 m
Atlantic Ocean	9000 m
Indian Ocean	7000 m
Arctic Ocean	6000 m

(b) Atlantic Ocean

Test yourself (p 14)

T1 (a) 17 252 (b) 5400 (c) 4000
T2 (a) 2760 (b) 2800
T3 (a) 6580
(b) (i) Four thousand eight hundred and seventy-three
(ii) 4900
(c) 5107, 7105, 7150, 7510

3 Shapes

A Circles (p 15)

A1 (a) The orange line and the blue line
(b) 4 cm
(c) 2 cm
A2 About 60 cm
A3 (a) (i) E (ii) C
(b) (i) 4 cm or 40 mm (ii) 1.5 cm or 15 mm
A4 (a) A circle of diameter 8 cm
(b) A circle of diameter 10 cm

B Triangles (p 16)

B1 A and D
B2 2.4 cm, 6.5 cm, 6.5 cm; yes, it is an isosceles triangle.
B3 A, C and D
B4 A and D
B5

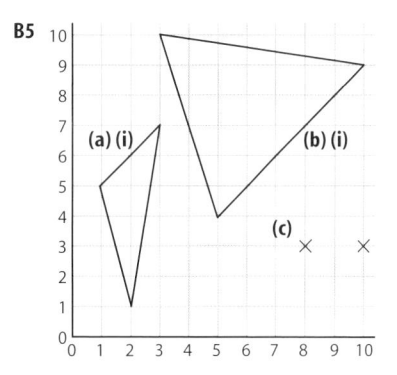

(a) (ii) Scalene
(b) (ii) Isosceles
(c) One of the following points: any point with 9 as the x-coordinate such as (9, 1) or (9, 7); (8, 1); (10, 1); (8, 5); (10, 5)

B6 (a) Isosceles
(b)

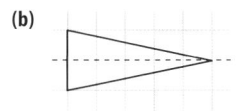

C Quadrilaterals (p 18)

C1 (a) A and D
(b) B (or A or D)
(c) C (or A or D, which are a special rhombus)

C2

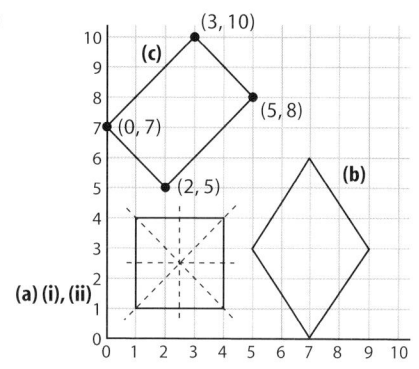

(a) (i), (ii) — square with centre around (3, 2.5) with diagonals and vertical/horizontal lines of symmetry shown
(c) square with vertices (0, 7), (3, 10), (5, 8), (2, 5)
(b) rhombus/kite with vertex (5, 8) area — (b) shape shown on grid

(b) (9, 3)

C3 (a) A, C and E (b) B and D

C4 (a) 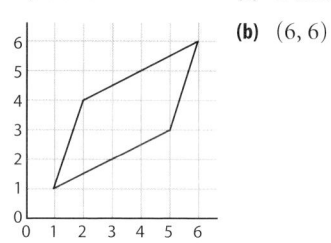 parallelogram drawn on grid (b) (6, 6)

C5 (a) (i) 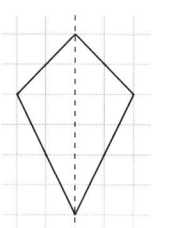 kite drawn (ii) Kite

(b) (i) 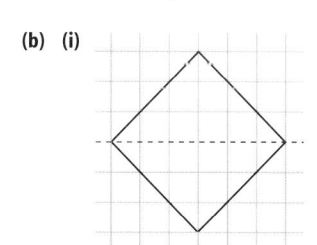 square (drawn as diamond) with horizontal line of symmetry
(ii) Square (it is also a rhombus but 'square' is the most informative description)

(c) (i) 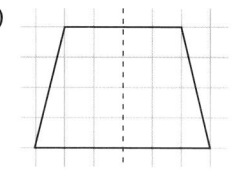 trapezium drawn (ii) Trapezium

D More than four edges (p 20)

D1 (a) 6 (the black and grey shapes are all pentagons)
(b) Trapeziums

D2 (a) 1   (b) Black   (c) Kites

D3 (a) (i) C, D and E   (ii) B and F
(b) (i) G   (ii) A   (iii) C

D4 (a) Hexagon   (b) 2

D5 (a) Any five-sided shape with one line of symmetry
(b) The line of symmetry shown on the shape

D6 (a) (i) 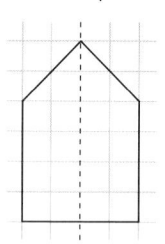 pentagon (house shape) drawn   (ii) Pentagon

(b) (i) 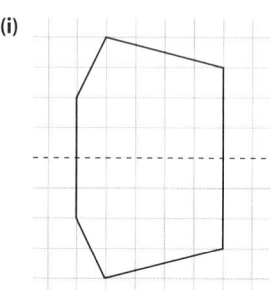 hexagon drawn with horizontal line of symmetry   (ii) Hexagon

(c) (i) 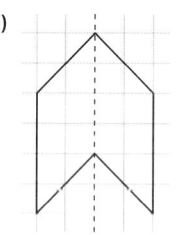 hexagon (arrow shape) drawn   (ii) Hexagon

E Shading squares to give reflection symmetry (p 22)

E1 shaded grid with horizontal and vertical lines of symmetry

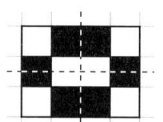

E2 shaded grid

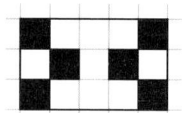

E3 Any pattern with 8 squares shaded including these

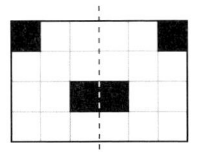

that also has one vertical line of symmetry, for example:

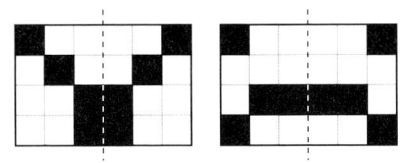

E4 The different ways (not including reflections or rotations) are:

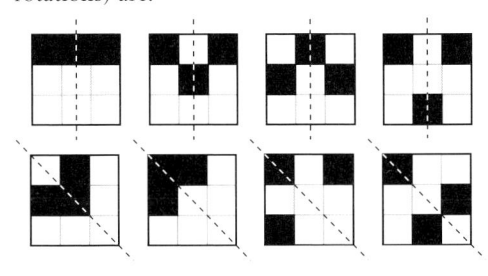

Test yourself (p 23)

T1 About 65 to 67.5 metres

T2 (a)   (b) Isosceles

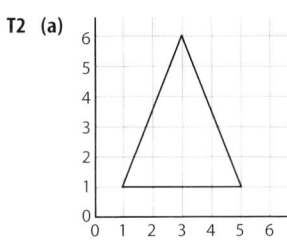

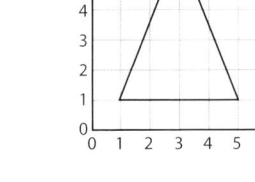

T3 (a) Pentagon   (b) 6

T4 (a), (b)

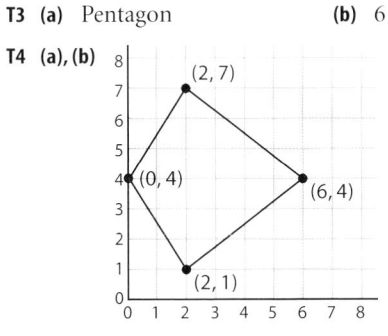

T5

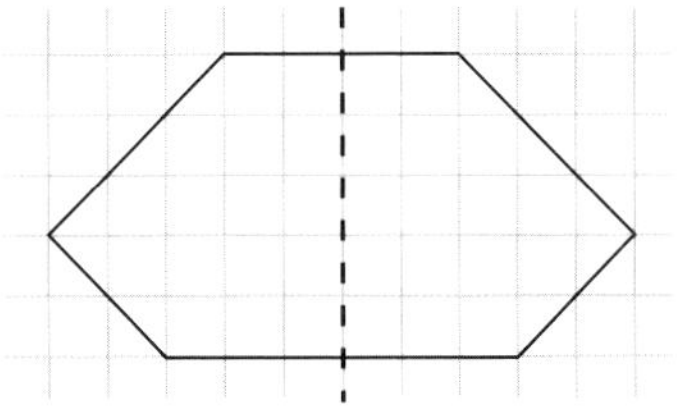

T6 (a) 4 cm

(b), (c) A circle of diameter 4 cm drawn on the sheet.

T7 (a), (b), (c)

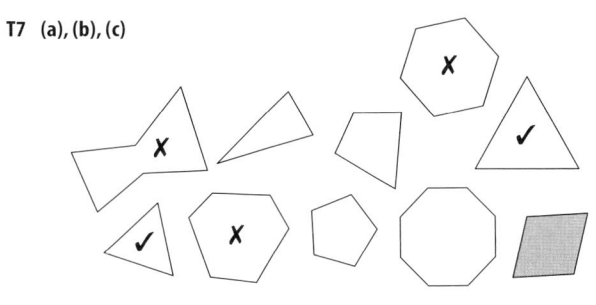

T8 (a) Octagon

(b)

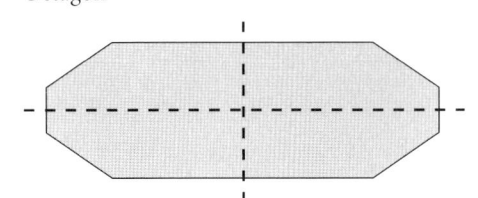

4 Adding and subtracting whole numbers

A Using written and mental methods (p 24)

A1 (a) 24   (b) 10

A2 (a)

22	2	15
6	13	**20**
11	**24**	4

(b)

12	**17**	**18**
15	19	23
20	**21**	16

(c)

30	**48**	36
44	38	**32**
40	**28**	46

(d)

70	**106**	82
98	**86**	74
90	**66**	102

A3 (a) 57   (b) 82   (c) 113
    (d) 692   (e) 671   (f) 325

A4 (a) 26   (b) 8   (c) 47
    (d) 125   (e) 316   (f) 165

A5 Puzzle 1

113	146	126	101
181	111	95	115
132	30	136	51
164	75	211	134

Total 390

Puzzle 2

9	7	21	76
65	92	79	33
26	23	3	94
44	72	58	36

Total 161

Puzzle 3

284	185	773	1313
468	175	428	2612
682	702	1243	187
1023	28	834	25

Total 2555

A6 (a) A: 33 kg, B: 29 kg, C: 25 kg, D: 33 kg, E: 29 kg
    (b) C: 383 kg

A7 (a) 184   (b) 69

A8 (a)
```
  6 1
+ 2 7
-----
  8 8
```
(b)
```
  5 7
+ 3 4
-----
  9 1
```
(c)
```
  8 7
+ 3 9
-----
1 2 6
```
(d)
```
  7 6
- 4 8
-----
  2 8
```
(e)
```
  4 2
- 1 9
-----
  2 3
```

A9 (a) (i) 12 and 47   (ii) 25 and 9
     (iii) 56 and 71   (iv) 35 and 68
  (b) (i) 25 and 12   (ii) 47 and 56

Test yourself (p 25)

T1 (a) 71   (b) 122   (c) 38   (d) 58

T2 126 m

T3 208

5 Listing

A Arrangements (p 26)

A1

First	Second	Third
●	▲	■
●	■	▲
▲	●	■
▲	■	●
■	●	▲
■	▲	●

A2 (a)

First	Second	Third
A	E	P
A	P	E
E	A	P
E	P	A
P	A	E
P	E	A

(b) Two ways spell a word, APE and PEA.

A3 (a)

First	Second	Third
8	3	5
8	5	3
5	3	8
5	8	3
3	8	5
3	5	8

(b) 853
(c) 358

B Combined choices (p 27)

B1

Drink	Snack
Tea	Muffin
Tea	Cookie
Tea	Bagel
Coffee	Muffin
Coffee	Cookie
Coffee	Bagel
Hot chocolate	Muffin
Hot chocolate	Cookie
Hot chocolate	Bagel

B2

Joe	Naz
Football	Football
Football	Athletics
Football	Cricket
Athletics	Football
Athletics	Athletics
Athletics	Cricket
Cricket	Football
Cricket	Athletics
Cricket	Cricket

B3 (a)

First	Second	Third
E	M	H
E	H	M
M	E	H
M	H	E
H	E	M
H	M	E

(b) There are two ways with English last.

Test yourself (p 28)

T1 (a)

5	4	6
5	6	4
4	5	6
4	6	5
6	5	4
6	4	5

(b) 4
(c) 654

6 Multiplying and dividing whole numbers

A Multiplying whole numbers (p 29)

A1 (a) 84 (b) 117 (c) 108 (d) 160
(e) 138 (f) 248 (g) 370 (h) 273

A2 (a) $35 \times 2 = 70$
(b) (i) $25 \times 3 = 75$ (ii) $23 \times 5 = 115$
(iii) $53 \times 2 = 106$
(c) $32 \times 5 = 160$ is the largest result.

A3 140p or £1.40

A4 168

A5 312

A6 112

B Dividing whole numbers (p 30)

B1 (a) 19 (b) 14 (c) 17 (d) 17
(e) 19 (f) 15 (g) 13 (h) 12

B2 14

B3 12

B4 (a) 11 remainder 1 (b) 28 remainder 1
(c) 14 remainder 3 (d) 12 remainder 1
(e) 11 remainder 4 (f) 13 remainder 5
(g) 10 remainder 3 (h) 28 remainder 2

B5 15

B6 21

B7 13

Test yourself (p 30)

T1 (a) 282 (b) 28 (c) 408 (d) 23

T2 (a) 216 (b) 12

T3 13

Review 1 (p 31)

1 (a) 21, 28 (b) 1, 2, 4, 8

2 900 km

3 8

4 (a) 407, 470, 704, 740
(b) 470, 740
(c) (i) 740 (ii) 40 or 4 tens

5 (a) 8030
(b) (i) 8130 miles (ii) 8000 miles

6 (a) 241 (b) 180 (c) 678 (d) 29

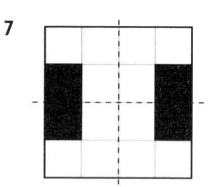

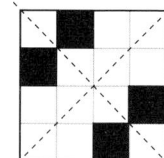

7 [diagram] or [diagram]

8 4006, 4060, 4600, 6004, 6400

9 96p

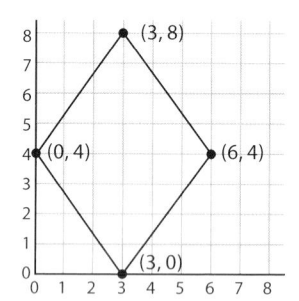

10 (a), (b), (c) [diagram with points (0,4), (3,8), (6,4), (3,0)]

(d) 2

7 Representing data

A Frequency charts and mode (for types of things) (p 32)

A1 (a) 12
(b) 11
(c) Japan
(d) [bar chart: Britain 12, France 8, Germany 11, Other European 7, Japan 13, Other 8]

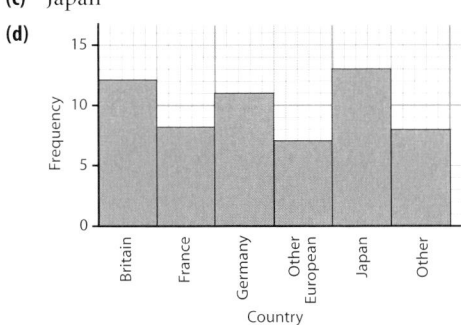

A2 (a)

Animal	Tally	Frequency
Cat	JHT JHT I	11
Dog	JHT III	8
Guinea pig	JHT	5
Hamster	IIII	4
Rabbit	III	3
Other	III	3

(b) [bar chart of above frequencies]
(c) Cat

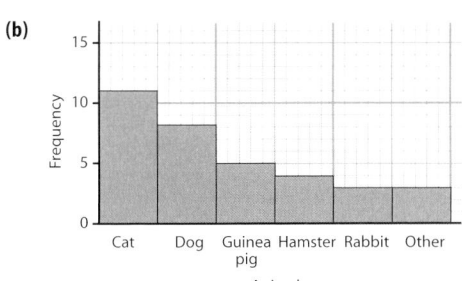

A3 (a) 12 (b) 16
(c) Cheese and onion (d) 78

A4 [bar chart: Europe 36, Asia 17, Africa 10, North America 48, South America 6, Australasia 15]

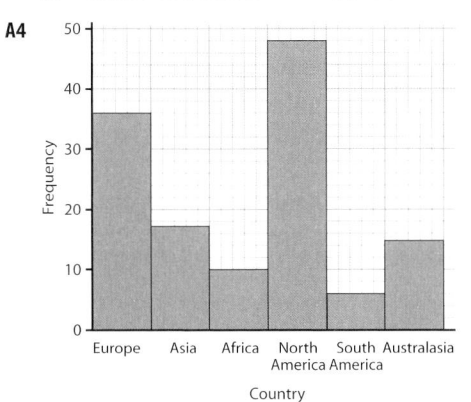

A5 [bar chart: Tea 94, White coffee 66, Black coffee 13, Hot chocolate 38, Herbal tea 10]

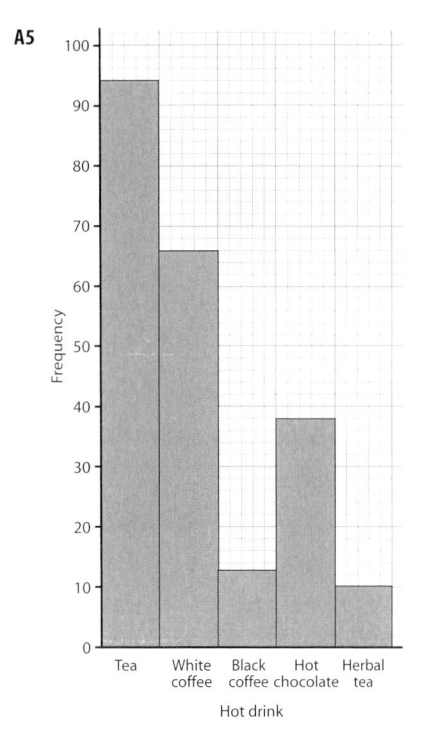

B Dual bar charts (p 34)

B1 (a) 14 (b) 20 (c) Black
(d) Brown (e) Maroon

B2 (a) 17 (b) City
(c) Rovers and Lions

B3 (a) Greenwich (b) Tower of London
(c) London Zoo

B4 [dual bar chart Male/Female: Cattle 15/33, Sheep 27/30, Pigs 9/15, Goats 7/4]

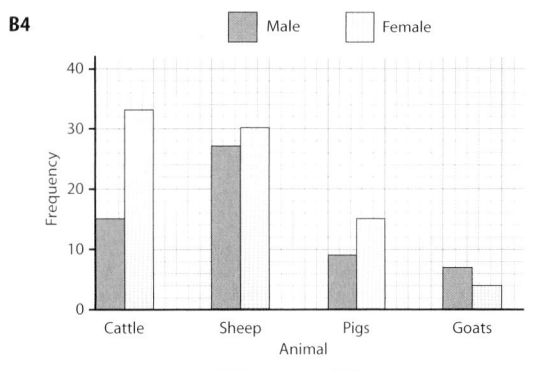

B5 [dual bar chart Boys/Girls: English 22/30, Maths 27/19, Sciences 12/10, History 7/23, Geography 11/8]

Answers: Chapter 7 191

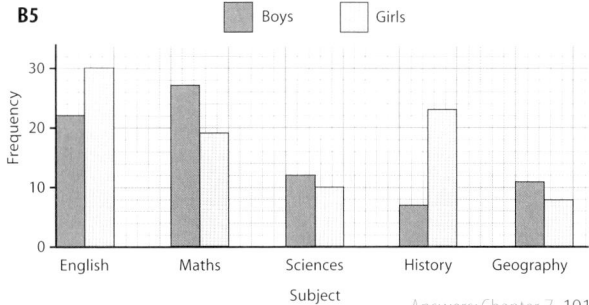

C Pictograms (p 35)

C1 (a) 50   (b) 5 trees   (c) 15   (d) 125

C2

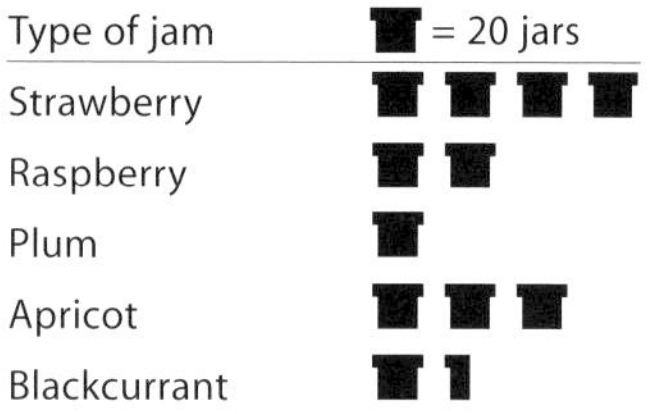

Type of jam	= 20 jars
Strawberry	▛ ▛ ▛ ▛
Raspberry	▛ ▛
Plum	▛
Apricot	▛ ▛ ▛
Blackcurrant	▛ ▍

C3

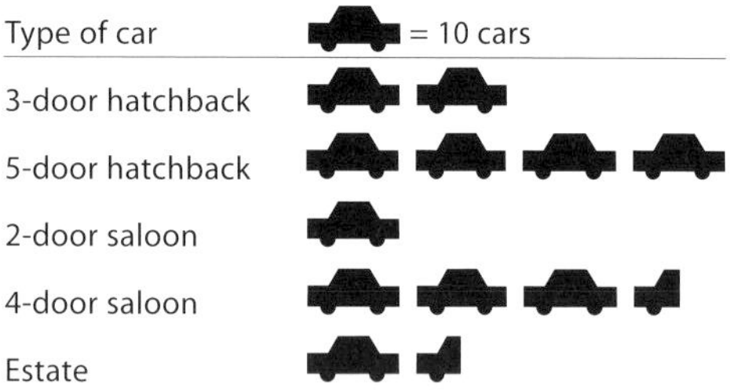

Type of car	= 10 cars
3-door hatchback	🚗 🚗
5-door hatchback	🚗 🚗 🚗 🚗
2-door saloon	🚗
4-door saloon	🚗 🚗 🚗 🚗
Estate	🚗 🚗

D Frequency charts and mode (for quantities) (p 36)

D1 (a)

Number of people in car	Tally	Frequency
1										10
2											11
3									8
4						4
5				2

(b) 10 + 11 + 8 + 4 + 2 = 35

D2 (a) 1   (b) 31

D3 (a)

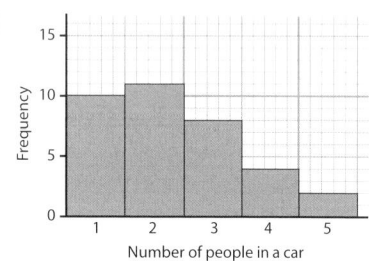

(b) 2 people in a car

D4 (a)

Number of eggs in a nest	Tally	Frequency
0											11
1					3
2							6
3						5
4								7
5					3

(b)

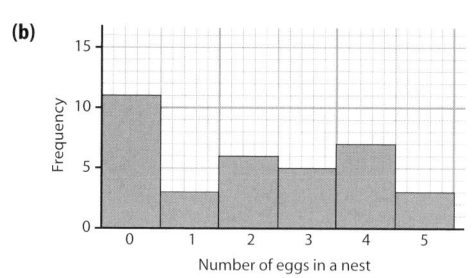

(c) 0 eggs in a nest

D5

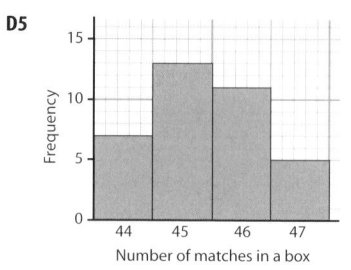

D6 (a)

Number of goals scored in a match	Tally	Frequency
0						5
1								7
2						5
3								7
4						4
5				2

(b)

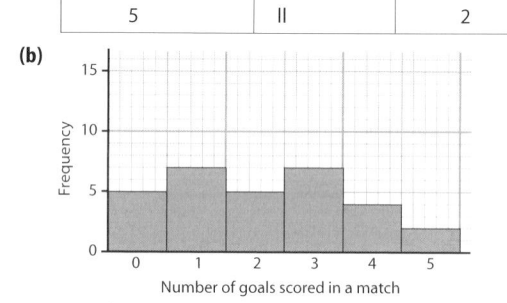

(c) 1 and 3 goals in a match

Test yourself (p 37)

T1 (a) Ready salted   (b) 36
     (c) 30   (d) 176

T2 (a)

Colour	Tally	Frequency
Red	|||| ||||		11
Yellow	|||| ||	7
Green	|||| |||	8
Orange	|||| |||	8

(b) Red

(c)

[Bar chart: Frequency vs Colour — Red ~11, Yellow ~7, Green ~8, Orange ~8]

T3 (a) 24 **(b)** Tuesday
(c) Friday **(d)** 6
(e) Yes; there are 16 boys and 8 girls; 16 = 8 × 2

T4 (a) Jack **(b)** 15

8 Fractions

A Recognising fractions (p 39)

A1 (a) Yes **(b)** No, $\frac{1}{8}$ **(c)** No, $\frac{1}{2}$ **(d)** Yes

A2 (a) $\frac{1}{3}$ **(b)** $\frac{1}{4}$ **(c)** $\frac{5}{6}$ **(d)** $\frac{3}{5}$
(e) $\frac{2}{5}$ **(f)** $\frac{1}{2}$ **(g)** $\frac{1}{4}$ **(h)** $\frac{3}{4}$

A3 6 by 2 rectangle with 3 squares shaded

A4 (a) 6 by 2 rectangle with 9 squares shaded
(b) 6 by 2 rectangle with 4 squares shaded
(c) 6 by 2 rectangle with 8 squares shaded
(d) 6 by 2 rectangle with 2 squares shaded
(e) 6 by 2 rectangle with 10 squares shaded

B Finding a fraction of a number (p 40)

B1 (a) To find $\frac{1}{2}$ of a number divide it by **2**.
(b) (i) 15 **(ii)** 18 **(iii)** 34
(iv) 140 **(v)** 75

B2 (a) To find $\frac{1}{4}$ of a number divide it by **4**.
(b) (i) 6 **(ii)** 11 **(iii)** 25
(iv) 30 **(v)** 45

B3 (a) 5 **(b)** 4 **(c)** 2 **(d)** 9 **(e)** 3

B4 (a) 30 **(b)** 15 **(c)** 20

B5 (a) (i) 6 **(ii)** 12
(b) (i) 10 **(ii)** 50

B6 (a) 27 **(b)** 16 **(c)** 18 **(d)** 25 **(e)** 8
(f) 120 **(g)** 240 **(h)** 120 **(i)** 45 **(j)** 90

B7 (a) TRICKS **(b)** PRECIOUS
(c) DISAPPEAR **(d)** THIRSTY
(e) COMPANION

B8 (a) $\frac{1}{3}$ of **24** = 8 **(b)** $\frac{1}{4}$ of 12 = 3
(c) $\frac{1}{8}$ of 40 = 5 **(d)** $\frac{3}{4}$ of **24** = 18

Test yourself (p 41)

T1 (a) $\frac{6}{10}$ or $\frac{3}{5}$
(b) 4 by 3 rectangle with 9 squares shaded

T2 (a) 9 **(b)** 5 **(c)** 15 **(d)** 30 **(e)** 90

9 Decimal places

A One decimal place (p 42)

A1 (a) 5.1 (b) 5.8 (c) 10.2 (d) 10.5 (e) 0.3 (f) 0.7

A2 (a) The numbers 3.1, 3.4, 3.6, 3.8, 4, 4.5 and 4.9 marked on the number line to make the word DECAGON
(b) 8.0, 8.6, 9.1, 9.7, 10.2, 10.9, 11.3; PYRAMID
(c) 4.9, 5.1, 5.8, 6.3, 6.9, 7.3; HEIGHT
(d) 0.1, 0.3, 0.7, 1.5, 1.9, 2.6, 3.0, 3.6; CYLINDER

A3 (a) 9.9 (b) 9.3 (c) 2.1 (d) 2.5

A4 (a) True (b) True (c) True (d) False

A5 A and C

A6 3.3, 2.3, 3, 3.8

A7 (a) 1.5, 3.2, 4.9, 5.4, 10.1
(b) 1.2, 2, 3.4, 6.7, 7
(c) 0.1, 0.9, 1.2, 1.7, 2
(d) 0.1, 0.2, 0.5, 1, 1.3

A8 (a) 6.9 cm (b) 7.1 cm

A9 2.5

A10 (a) (i) 5.4 cm (ii) 4.9 cm (iii) 4.6 cm (iv) 4.8 cm
(b) February, March, April
(c) 6.4 cm

B Two decimal places (p 44)

B1 (a) 5.23 (b) 5.27 (c) 2.82 (d) 2.91 (e) 2.96
(f) 3.01 (g) 3.06 (h) 0.02 (i) 0.09 (j) 0.15

B2 (a) True (b) False (c) True (d) True

B3 (a) The numbers in order are 2.51, 2.58, 2.6, 2.63, 2.68, 2.7 and 2.71, and the letters make the word DECIMAL.
(b) 7.34, 7.39, 7.4, 7.43, 7.5, 7.52; SQUARE
(c) 4.02, 4.1, 4.12, 4.18, 4.2, 4.23; RADIUS
(d) 0.03, 0.1, 0.16, 0.21, 0.29, 0.3; CUBOID

B4 (a) R and U (b) P, Q, R, S and U

B5 (a) The numbers in order are 2.28, 2.34, 2.37 and 2.41, and the letters make the word MEAN.
(b) 1.1, 1.16, 1.2, 1.29, 1.3; PRISM
(c) 4.97, 5, 5.09, 5.1, 5.14; ANGLE
(d) 0.02, 0.1, 0.13, 0.2, 0.27; RANGE

B6 2.6, 2.67, 2.52

B7 7.1, 7.28, 7.4, 7.92, 8

B8 1.2, 1.03, 0.3, 0.06

B9 3.9 kg

B10 (a) 3.5 (b) 3.25 (c) 9.75 (d) 1.05

B11

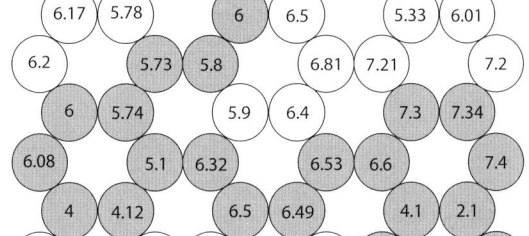

You finish on 4.1.

C Decimal lengths (p 46)

C1 (a) 4.2 metres (b) 6.5 metres

C2 (a) 1.1 metres (b) 2.3 metres (c) Yes

C3 (a) 3 metres (b) 3.6 metres (c) Yes

C4 (a) 1.1 metres
(b) The carpets with width 1.5 m, 1.15 m and 1.25 m

C5 Yes

C6 No

Test yourself (p 47)

T1 (a) 2.3 (b) 2.9

T2 Three different numbers between 4 and 5

T3 A

T4 3.3

T5 2.09, 2.1, 2.36, 2.7

10 Median and range

A Median (for an odd number of data items) (p 48)

A1 (a) 53 54 62 64 65 70 75
(b) 64 g

A2 (a) 44 46 49 50 51; median 49
(b) 28 29 33 37 44 47 50 51 63; median 44
(c) 38 43 43 51 55 63 65; median 51
(d) 2 3 3 4 5 5 6 6 7 7 9; median 5

B Median (for any number of data items) (p 49)

B1 58 62; median 60 g

B2 (a) 54 59 63 67 71 80
(b) 63 67; median 65 g

B3 53 56; median 54.5 g

B4 (a) 58 62 64 67 67 71
(b) 65.5 g

B5 (a) 15 g   (b) 54 g   (c) 97.5 g   (d) 28.5 g

B6 (a) 60 g
(b) 62.5 g
(c) The median weight went up.

C Range (p 50)

C1 (a) (i) 89 g   (ii) 39 g   (iii) 50 g
(b) (i) 630 g   (ii) 480 g   (iii) 150 g
(c) (i) 82 g   (ii) 48 g   (iii) 34 g
(d) (i) 673 g   (ii) 504 g   (iii) 169 g

C2 (a) 159 162 166 168 171 174 177
(b) 168 cm
(c) 18 cm

C3 (a) 165 cm   (b) 9 cm

C4 (a) Median 60, range 34
(b) Median 65, range 30
(c) Median 41, range 40

C5 (a) 23.5   (b) 29   (c) 25   (d) 31

D Comparing two sets of data (p 51)

D1 (a) 24   (b) 27   (c) Team B
(d) 13 years   (e) 15 years   (f) Team B

D2 (a) (i) 38 kg   (ii) 11 kg
(b) (i) 39 kg   (ii) 14 kg
(c) Boys
(d) Boys

D3 (a) (i) 20.5 s   (ii) 12 s
(b) (i) 41.5 s   (ii) 15 s
(c) Left-hand times

D4 (a) (i) 27 s   (ii) 12 s
(b) Boys
(c) Girls

Test yourself (p 52)

T1 (a) 53   (b) 27

T2 21

T3 (a) 14.5 °C   (b) 5 degrees   (c) 13 °C
(d) 6 degrees   (e) Yorkshire   (f) Wales

11 Mental methods 2

A Multiplying by 10, 100, 1000 (p 53)

A1 (a) 50 (b) 500 (c) 5000 (d) 25 (e) 250
A2 (a) 250 (b) 43 200 (c) 6500 (d) 72 000 (e) 80 000
(f) 24.5 (g) 320 (h) 4210 (i) 200 (j) 58
A3 325 g
A4 (a) 17 mm (b) 17.8 mm

B Dividing by 10, 100, 1000 (p 54)

B1 (a) 40 (b) 4 (c) 0.4 (d) 6.2 (e) 0.62
B2 (a) 1.8 (b) 56.1 (c) 3.4 (d) 7.64 (e) 4.3
(f) 0.25 (g) 0.36 (h) 0.45 (i) 0.97 (j) 0.03
B3 £0.45 or 45p
B4 2.75 g
B5 (a) £0.14 or 14p (b) 130 mm

C Multiplying by numbers ending in zeros (p 54)

C1 (a) 160 (b) 800 (c) 2100 (d) 240 (e) 450
(f) 150 (g) 540 (h) 1600 (i) 2700 (j) 140
C2 (a) 120 (b) 1200 (c) 12 000
(d) 12 000 (e) 120 000
C3 (a) 18 000 (b) 140 000 (c) 18 000
(d) 3600 (e) 15 000
C4 (a) 20 000 (b) 4800 (c) 35 000
(d) 80 000 (e) 300 000
C5 £60
C6 (a) £32 (b) £100 (c) £240
C7 £80 000

Test yourself (p 55)

T1 (a) 920 (b) 31 000 (c) 405 (d) 260 (e) 68
(f) 42 (g) 80 (h) 0.79 (i) 0.43 (j) 0.05
T2 (a) 1800 (b) 8000 (c) 400 000
(d) 36 000 (e) 420 000
T3 (a) £5.60 (b) £1.60

12 Solids, nets and views

A Solids and nets (p 56)

A1 (a) (i) Cuboid (ii) 6 faces, 8 vertices, 12 edges
(b) (i) Pyramid (ii) 5 faces, 5 vertices, 8 edges
(c) (i) Prism (ii) 5 faces, 6 vertices, 9 edges
A2 W and X
A3 (a) Cuboid (b) Prism (c) Pyramid
A4 (a) Any one of these diagrams:

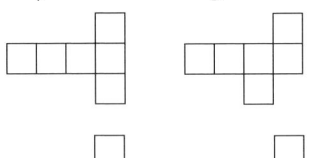

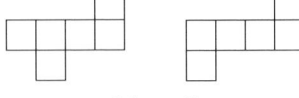

(b) Any one of these diagrams:

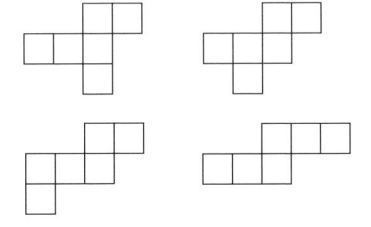

(c) Any one of these diagrams:

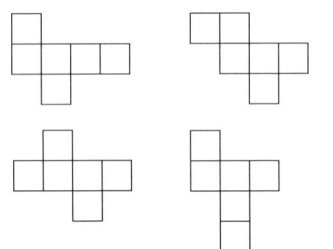

A5 (a) Prism (b) 6 faces (c) 8 vertices

B Views (p 58)

B1 (a) R (b) Q (c) S (d) P (e) T
B2 A: Triangle B: Star C: Flower D: Heart
E: Star F: Triangle G: Square H: Star
B3 X

Test yourself (p 59)

T1 (a) Sphere (b) Cylinder (c) Pyramid
T2 (a) Yes (b) Yes (c) No (d) Yes
T3 (a) Prism (b) 5 faces (c) 6 vertices
T4 P

13 Weighing

A Grams and kilograms (p 60)

A1 (a) grams (b) kilograms (c) grams

A2 2000 grams

A3 (a) 3000 g (b) 9000 g (c) 10 000 g
(d) 500 g (e) 1250 g

A4 850 g

A5 (a) 2 (b) 4 (c) 10 (d) 5
(e) 20 (f) 50 (g) 8 (h) 100

A6 4 kg

B Using decimals (p 61)

B1 (a) 500 g (b) 3500 g (c) 400 g (d) 750 g
(e) 3200 g (f) 4250 g (g) 1360 g (f) 1450 g

B2 4.5 kg ripe bananas
0.5 kg chopped nuts
1 kg soya margarine
1.25 kg raisins
0.75 kg rolled oats
1.5 kg wholewheat flour

B3 1700 g or 1.7 kg

B4 7 g, 200 g, 0.7 kg, 1200 g, 1.5 kg

Test yourself (p 61)

T1 6000 g

T2 450 g or 0.45 kg

T3 (a) 2 kg (b) 1.25 kg (c) 0.65 kg

T4 1600 g

T5 1.5 kg

Review 2 (p 62)

1 (a) 4
(b) 6
(c) Blue tit
(d)

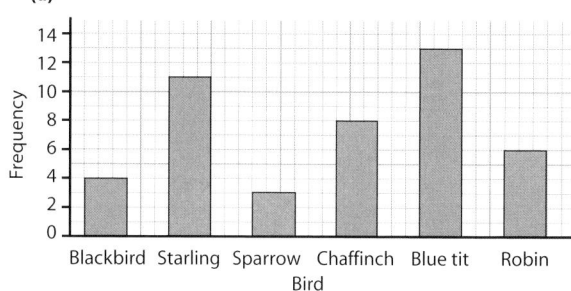

(e) 45

2 36

3 (a) 31 2 or 31.5 years (b) 17 years

4 (a) 320 (b) 89.1 (c) 50 (d) 6.5 (e) 35 000

5 (a) 1.25 kg (b) 250 grams

6 Pyramid

7 800 g, 1250 g, 12 kg, 1.9 kg

8 X

9 1.08, 1.3, 3.58, 3.6, 4

14 Time and travel

A Understanding 12-hour and 24-hour clock time (p 63)

A1 (a) 15:40 (b) 00:30 (c) 01:40
(d) 21:15 (e) 08:40 (f) 11:50

A2 (a) 9:00 a.m. (b) 12:30 p.m. (c) 4:45 p.m.
(d) 8:10 p.m. (e) 11:40 p.m.

A3 03:05, 7:45 a.m., 2:00 p.m., 17:35, 7:15 p.m., 23:15

B Time intervals (p 64)

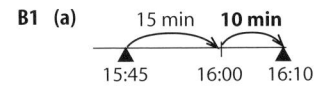

B1 (a)

```
      15 min    10 min
    ▲         ▲        ▲
  15:45     16:00    16:10
```

(b) 25 minutes

B2 (a) 10 minutes (b) 45 minutes (c) 15 minutes
(d) 45 minutes (e) 55 minutes (f) 45 minutes

B3 (a) 7:50 a.m. (b) 8:35 a.m.

B4 30 minutes

B5 4:25 p.m.

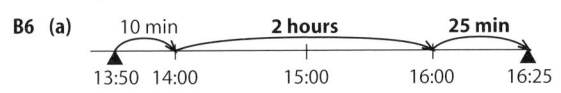

B6 (a)

```
     10 min       2 hours         25 min
   ▲       ▲                ▲           ▲
 13:50  14:00            15:00  16:00  16:25
```

(b) 2 hours 35 minutes

B7 (a) 3 hours 15 minutes (b) 2 hours 30 minutes
(c) 4 hours 30 minutes (d) 1 hour 45 minutes
(e) 1 hour 45 minutes (f) 2 hours 50 minutes

B8 9:20 a.m.

C Working out starting times (p 65)

C1 (a) 11:00 (b) 3:20 p.m. (c) 10:30 a.m.

C2 (a) 8:20 a.m. (b) 16:35 (c) 9:25 a.m.

C3 (a) (i)

```
       15 min   10 min
     ▲        ▲       ▲
   08:45    09:00   09:10
```

(ii) 08:45

(b) (i) 10:45 a.m. (ii) 19:45
(iii) 7:35 p.m. (iv) 03:35

C4 7:40 a.m.

C5 4:40 p.m.

C6 (a) 9:30 a.m. (b) 7:00 a.m.

D Timetables (p 66)

D1 Six

D2 0926

D3 1 hour 56 minutes

D4 (a) 1 hour 46 minutes (b) 10 minutes

D5 The 0817 train from Diss

D6 The 0807 train from Ipswich

D7 17 minutes

Test yourself (p 67)

T1 (a) 8:10 (b) 2 hours 35 minutes

T2 (a) 9:45 (b) 47 minutes

T3 (a) 0930 (b) 2 hours 45 minutes
(c) 17 minutes

15 Angle

A Drawing, measuring and sorting angles (p 68)

A1 180

A2 90

A3 It is a square: all the sides are the same and all the angles are 90°.

A4 B and C

A5 C, A, B (20°, 60°, 80°)

A6 Angles drawn as specified

A7 A and D

A8 A, C, B (110°, 125°, 170°)

A9 Angles drawn as specified

A10 B and D

A11 B, A, C (260°, 300°, 340°)

A12 Angles drawn as specified

A13 (a) Obtuse (b) Reflex (c) Right angle (d) Reflex

A14 Angle *a* is a right angle.
Angle *d* is obtuse.
Angle *b* is reflex.
Angle *c* is acute.

B Turning (p 71)

B1 Right angle

B2 (a) Obtuse (b) Acute (c) Reflex (d) Acute (e) Half turn

B3 (a) 70° (b) Acute

B4 (a) 120°, obtuse (b) 90°, right angle (c) 190°, reflex (d) 120°, obtuse

B5 Anticlockwise

C Angles in shapes (p 72)

C1 (a) Isosceles (two sides the same)
(b) Two of the angles are the same (78°).

C2 (a) Equilateral (all three sides the same)
(b) They are all the same (60°).

C3 (a) Opposite sides are equal; opposite angles are the same (55° and 125°).
(b) Parallelogram

C4 (a) An accurate drawing as specified
(b) (i) 31° (ii) 59° (iii) 11.7 cm

C5 (a) An accurate drawing as specified
(b) QR = 6.3 cm, PR = 11.0 cm

D Estimating angles (p 73)

D1 (a) Less (b) More (c) Less (d) More

D2 (a) 20°, 30°, or 40° (b) 60° or 70° (c) 10° or 20° (d) 60° or 70°

D3 (a) More (b) Less (c) Less (d) More

D4 (a) 290° or 300° (b) 240° or 250° (c) 250° or 260° (d) 320° or 330°

D5 (a) Less (b) Less (c) More (d) More

D6 (a) 100° or 110° (b) 110° or 120° (c) 140° or 150° (d) 150°, 160° or 170°

D7 B

D8 C

Test yourself (p 74)

T1 (a) 30° (b) Acute

T2 130°

T3 (a) 45° (b) Acute

T4 (a) 135°, obtuse (b) 270°, reflex (c) 225°, reflex (d) 180°, half turn

T5 Anticlockwise

T6 (a) An accurate drawing as specified
(b) 7.2 cm

T7 (a) 40° or 50° (b) 70° or 80° (c) 200° or 210°

16 Length

A Centimetres and millimetres (p 75)

A1 (a) (i) 30 mm (ii) 3 cm
   (b) (i) 10 mm (ii) 1 cm
   (c) (i) 25 mm (ii) 2.5 cm
   (d) (i) 53 mm (ii) 5.3 cm
   (e) (i) 7 mm   (ii) 0.7 cm
   (f) (i) 127 mm (ii) 12.7 cm

A2 A straight line 5.6 cm long

A3 A rectangle with a length of 6.4 cm and a width of 3.8 cm

A4 11 cm

A5 (a) mm (b) cm (c) cm (d) mm

A6

Spider	Span (mm)	Span (cm)
Bird-eating Spider	250 mm	**25 cm**
Tarantula	**240 mm**	24 cm
Raft Spider (UK)	145 mm	**14.5 cm**
House Spider (UK)	**75 mm**	7.5 cm
Wolf Spider (UK)	17 mm	**1.7 cm**
Money Spider (UK)	3 mm	**0.3 cm**

A7 5 mm, $2\frac{1}{2}$ cm, 2.7 cm, 33 mm, 5 cm

A8 162.3 cm

B Metres and centimetres (p 76)

B1 400 cm

B2 (a) 200 cm (b) 1000 cm (c) 450 cm (d) 25 cm

B3 (a) cm (b) m (c) cm (d) m

B4
Python   600 cm
Anaconda   850 cm
Grass Snake 125 cm
Adder   75 cm

B5 2 m

B6
Fruit Bat   1.7 m
Mouse-eared Bat (UK)   0.45 m
Pipistrelle Bat (UK)   0.25 m
Kitti's Hog-nosed Bat   0.09 m

B7 1.4 m or 140 cm

B8 1.25 m or 125 cm

B9 (a) 3 cm, 30 cm, 3 m, 30 m
   (b) 76 cm, 1.36 m, 150 cm, 2.3 m
   (c) 15 cm, 105 cm, 1.2 m, $1\frac{1}{4}$ m
   (d) 5 mm, 50 mm, 50 cm, 5 m

C Kilometres (p 77)

C1 (a) 5000 m (b) 2000 m (c) 500 m (d) 750 m
C2 (a) 4 km   (b) 3 km   (c) 20 km (d) 100 km
C3 (a) 2500 m (b) 9400 m (c) 600 m (d) 250 m
C4 (a) 1.5 km (b) 5.8 km (c) 0.3 km (d) 2.64 km
C5 14.2 km or 14 200 m

Test yourself (p 77)

T1 (a) 9 cm
   (b) (i) A straight line 7 cm long
       (ii) A point marked with a cross in the middle of the line (3.5 cm from either end)

T2 (a) 7.2 cm (b) 540 cm

T3 2 mm, 20 mm, 20 cm, 2 m

T4 700 m

17 Squares and square roots

A Square numbers and square roots (p 78)

A1 (3×3 grid of dots)

A2 (a) Yes (b) 4
A3 (a) 49 (b) 7 (c) 1
A4 (a) 4 (b) 25 (c) 81
A5 1, 4, 9, 16, 25, 36, 49, 64, 81, 100, 121, 144
A6 11
A7 (a) 3 (b) 5 (c) 10 (d) 6 (e) 8

B Using shorthand (p 79)

B1 (a) 9 (b) 100 (c) 64
B2 $10^2 = 10 \times 10 = \mathbf{100}$
B3 $3^2 = 9$; $7^2 = 49$; $1^2 = 1$; $8^2 = 64$
B4 (a) 25 (b) 36 (c) 4 (d) 81 (e) 144
B5 (a) 5 (b) 2 (c) 1 (d) 9 (e) 10
B6 (a) 3 (b) 36 (c) 5 (d) 16

C Using a calculator (p 80)

C1 (a) 25 (b) 121 (c) 361 (d) 16 (e) 169 (f) 625
C2 (a) 8 (b) 14 (c) 7 (d) 18 (e) 20 (f) 31
C3 12
C4 225, 256, 289
C5 729

Test yourself (p 80)

T1 1, 4, 9, 25, 81
T2 (a) 81 (b) 25 (c) 64 (d) 7 (e) 3 (f) 6
T3 (a) 289 (b) 441 (c) 900 (d) 13 (e) 16 (f) 100
T4 1024

18 Adding and subtracting decimals

A Adding decimals (p 81)

A1 (a) 0.8 (b) 7.5 (c) 7.3 (d) 5.4
  (e) 9.9 (f) 3.6 (g) 1.5 (h) 6
A2 (a) £1.40 (b) £3.60 (c) £3.70 (d) £16.20
A3 £3.60
A4 (a) £4.47 (b) £6.82 (c) £30.38 (d) £52.05
A5 4.02 m
A6 (a) 6.2 m (b) 3.81 m (c) 20.64 m
A7 (a) 8.4 (b) 12.1 (c) 20 (d) 115.2
A8 (a) 1.2 and 1.3 (b) 0.3 and 1.3
  (c) 1.56 and 0.3 (d) 1.44 and 1.2
  (e) 1.56 and 1.3 (f) 1.44 and 1.56
A9 (a) 4.75 (b) 13.37 (c) 25.54 (d) 40.25

B Subtracting decimals (p 82)

B1 (a) 3 (b) 6.3 (c) 4.4 (d) 5.4
  (e) 1.9 (f) 0.3 (g) 7.5 (h) 0.8
B2 (a) £0.80 (b) £2.50 (c) £1.10 (d) £1.60
B3 (a) £3.70 (b) £3.18 (c) £8.68 (d) £0.31
B4 2.1 m
B5 £6.56
B6 (a) 13.1 (b) 14.83 (c) 1.8 (d) 1.37
  (e) 1.83 (f) 1.19 (g) 4.86 (h) 7.71
B7 13.6 cm
B8 1.17 m
B9 (a) 2.37 (b) 2.18 (c) 4.01 (d) 11.75
  (e) 36.84 (f) 2.24
B10 £3.74
B11 (a) 2.88 (b) 1.17 (c) 0.29

C Mixed questions (p 83)

C1 (a) £4.80 (b) £4.15 (c) £9.08 (d) £1.75
C2 (a) A: 0.28 m or 28 cm, B: 0.45 m or 45 cm,
   C: 0.05 m or 5 cm, D: 0.15 m or 15 cm,
   E: 0.11 m or 11 cm, F: 0.15 m or 15 cm
  (b) C: 4.9 m
C3 3.48 m

C4 Puzzle 1

13.61	7.2	5.32	5.17
5.8	11	2.36	13
6	4.51	6.3	16.82
4.63	5.9	16.79	7.6

Total 42.59

Puzzle 2

4.25	5.12	4.3	4.91
3.68	3.3	3.58	5.6
1.38	5.19	1.27	6.6
3.1	3.48	2.61	3.85

Total 17.09

Puzzle 3

1.28	12	1.5	10
6.31	5.78	1.8	3.62
22.16	16.15	7.21	4.42
5.43	9.3	16.2	4.47

Total 18.74

Test yourself (p 83)

T1 (a) £10.70 (b) £2.85
T2 (a) 6.49 kg (b) 0.67 kg
T3 1.65 m
T4 (a) 0.7 (b) 11.85 (c) 6.7 (d) 3.44

19 Mental methods 3

A Multiplying and dividing by 4 and 5 (p 84)

A1 (a) 52 (b) 92 (c) 14 (d) 100 (e) 128
  (f) 72 (g) 23 (h) 136 (i) 27 (j) 105
A2 (a) 120 (b) 140 (c) 170 (d) 360 (e) 330
  (f) 90 (g) 175 (h) 315 (i) 130 (j) 115
A3 (a) 68 (b) 96 (c) 26 (d) 84 (e) 324
  (f) 132 (g) 104 (h) 27 (i) 43 (j) 208
A4 (a) 520 (b) 240 (c) 34 (d) 41 (e) 130
  (f) 392 (g) 144 (h) 408 (i) 405 (j) 55

Review 3 (p 85)

1 (a) 108 (b) 360 (c) 77 (d) 5.4 (e) 6.5
2 (a) 8:15 a.m. (b) 3:10 a.m. (c) 4:25 p.m.
3 (a) 60 mm, 60 mm, 112 mm
  (b) Isosceles
  (c) (i) Obtuse (ii) 140°
4 (a) $4\frac{1}{2}$ cm or 4.5 cm (b) 45 mm
5 (a) 3 (b) 10
6 (a) 5.5 (b) 3.43 (c) 17.12 (d) 5.41
7 1.8 m or 180 cm
8 (a) Yes (b) 6 minutes
9 5 mm, 50 mm, 50 cm, 5 m, 0.5 km
10 (a) 16, 25, 49 (b) 15, 25, 40

20 Number patterns

A Simple patterns (p 86)

A1 (a) 37
  (b) An explanation such as:
  'You add 6 to get from one number to the next and 31 + 6 = 37. So 37 is the next number.'

A2 (a) 19, 16
  (b) An explanation such as:
  'You subtract 3 to get from one number to the next. 22 − 3 = 19 and 19 − 3 = 16, so 19 and 16 are the next two numbers.'

A3 (a) 15, 17 (b) 30, 34 (c) 41, 44
  (d) 65, 72 (e) 7, 5 (f) 10, 6

A4 (a) 36 (b) 51
  (c) 41, 36, 81, 101

A5 (a) Either 13, 21, 29, 37, 45, 53
  or 53, 45, 37, 29, 21, 13
  (b) Either 'You add 8' or 'You subtract 8'

A6 (a) 14
  (b) (i) 4, 8, 12, 16, 20, 24 (ii) 28
    (iii) 80
  (c) 3, 7, 11, 15, 19, 23
  (d) (i) Grey (ii) Grey (iii) White
    (iv) Orange (v) Blue

B Further patterns (p 87)

B1 (a) 26
  (b) An explanation such as:
  'To get from one number to the next you add 1, add 2, add 3, add 4, add 5, and so on. 20 + 6 = 26, so 26 is the next number in the sequence.'

B2 (a) 32, 64 (b) 2, 1 (c) 21, 27
  (d) 25, 19 (e) 31, 42 (f) 37, 69

B3 (a) 1 + 2 + 3 = 6
    2 + 3 + 4 = 9
    3 + 4 + 5 = 12
    4 + 5 + 6 = 15
    5 + 6 + 7 = 18
  (b) (i) **10 + 11 + 12 = 33** (ii) **11 + 12 + 13 = 36**
    (iii) **18 + 19 + 20 = 57** (iv) **7 + 8 + 9 = 24**
  (c) An explanation such as:
  'All the totals are multiples of 3. Because 28 is not a multiple of 3 the line cannot be completed as a line in the pattern.'

B4 (a) 1110
  (b) 2 × 5 = 10
    22 × 5 = 110
    222 × 5 = **1110**
    2222 × 5 = 11 110
    22 222 × 5 = 111 110
  (c) An explanation such as:
  'In each multiplication the number of 2s will be the same as the number of 1s. In Leonie's calculation there are seven 2s but only six 1s.'

Test yourself (p 88)

T1 (a) 18
  (b) An explanation such as:
  'You add 4 to get from one number to the next and 14 + 4 = 18. So 18 is the next number.'

T2 (a) (i) 53
    (ii) An explanation such as:
    'You add 5 to get from one number to the next and 48 + 5 = 53. So 53 is the next number.'
  (b) An explanation such as:
  'All numbers in this pattern will end with a 3 or an 8. Because 200 ends in a 0, it is not in the pattern.'

T3 (a) 23, 27, 31, 35, 39 or 39, 35, 31, 27, 23
  (b) An explanation such as:
  'You add 4 to get from one number to the next and 39 + 4 = 43. So 43 is the next number.'
  or:
  'You subtract 4 to get from one number to the next and 23 − 4 = 19. So 19 is the next number.'

T4 48, 96

21 Estimating and scales

A Estimating lengths (p 89)

Estimates close to the answers given for A1 to A4 are acceptable.

A1 (a) 10 cm (b) 6 cm (c) 4 cm (d) 15 cm (e) 3 cm

A2 (a) 6 cm
(b) Length 4 cm, height 3 cm
(c) Length 8 cm, height 1 cm

A3 (a) 8 m (b) 16 m (c) 1.5 m (d) 2 m

A4 (a) 200 cm or 2 m (b) 75 cm (c) 50 cm

A5 (a) The policeman is about 2 metres tall.
This makes the Bargate about 12 metres high.
(b) An explanation of (a)

A6 (a) About 2 metres (b) About 1.7 metres

B Reading and estimating from scales (p 92)

B1 (a) 78 °F (b) 54 °F (c) 36 °F

B2 (a) 25 degrees
(b) (i) 250 °C (ii) 125 °C (iii) 275 °C

B3 (a) (i) 5 km/h (ii) 125 km/h
(b) (i) 2 m.p.h. (ii) 54 m.p.h.
(c) (i) 20 knots (ii) 440 knots

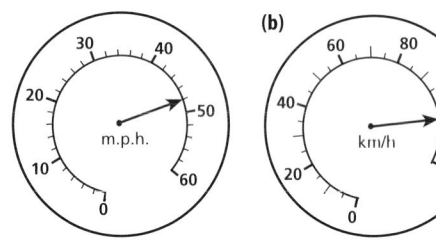

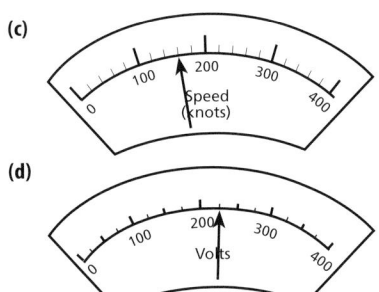

B4 (a), (b), (c), (d)

B5 (a) About 35 m.p.h. (b) About 45 m.p.h
(c) About 68 m.p.h.

C Decimal scales (p 93)

C1 Josie

C2 (a) 1.8 g (b) 5.4 cm (c) 18.6 °C

C3 (a) 1.5 (b) 2.5 (c) 6.25 (d) 3.85

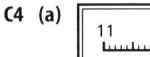

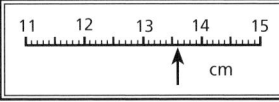

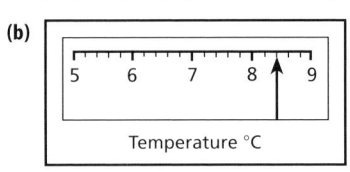

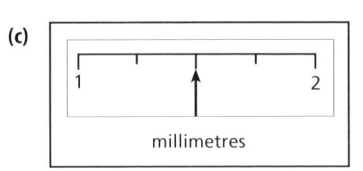

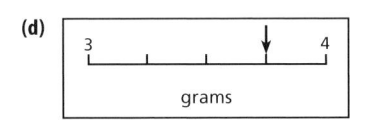

C4 (a), (b), (c), (d)

C5 (a) (i) 14.4 cm (ii) 15.5 cm (iii) 26.1 cm (iv) 13.0 cm
(b) January, October, December
(c) 28.5 cm
(d) 15.6 cm

C6 (a) About 7.4 (b) About 1.2
(c) About 0.7 (d) About 6.39

Test yourself (p 95)

T1 About 1.9 m or 2 m

T2 (a) 6 m
(b) (i) 34 m.p.h. (ii) 47 °C or 48 °C or 47.5 °C

T3 11.6 °C

22 Multiplying and dividing decimals

A Multiplying a decimal by a whole number (p 96)

A1 (a) 1.8 (b) 0.5 (c) 1.8 (d) 1.4
(e) 3.0 or 3 (f) 7.5 (g) 4.4 (h) 6.6

A2 A and D (1.2), B and E (2.4), C and G (1.6);
F (1.8) is the odd one out.

A3 (a) 21.5 (b) 16.2 (c) 22.2
(d) 13.3 (e) 44.8 (f) 19.5
(g) 39.2 (h) 45.0 or 45

A4 (a) **0.2** × 4 = 0.8 (b) 1.4 × **3** = 4.2
(c) 0.3 × **5** = 1.5 (d) **5** × 0.8 = 4
(e) **0.6** × 4 = 2.4 (f) 5 × **0.2** = 1

A5 15.0 kg or 15 kg

A6 17.6 cm

B Dividing a decimal by a whole number (p 97)

B1 (a) 4.2 (b) 3.1 (c) 6.2 (d) 7.1

B2 (a) 2.3 (b) 1.5 (c) 3.2 (d) 8.7

B3 (a) **2.4** ÷ 6 = 0.4 (b) 3 ÷ **5** = 0.6
(c) 3 ÷ **2** = 1.5 (d) **1.5** ÷ 3 = 0.5
(e) **2.4** ÷ 4 = 0.6 (f) 4.5 ÷ **5** = 0.9

B4 (a) 3.5 (b) 5.4 (c) 9.5 (d) 12.6

B5 A and F (8.2), B and D (5.7), C and G (6.4);
E (6.8) is the odd one out.

B6 0.5 m

Test yourself (p 97)

T1 (a) 3.5 (b) 4.2
(c) 22.8 (d) 60.0 or 60

T2 (a) 0.2 (b) 0.8 (c) 1.9 (d) 1.2

T3 10.5 litres

23 Area and perimeter

A Shapes on a grid of centimetre squares (p 98)

A1 (a) $2\,cm^2$ (b) $3\,cm^2$ (c) $5\,cm^2$ (d) $4\,cm^2$

A2 (a) $2\frac{1}{2}\,cm^2$ (b) $2\,cm^2$ (c) $4\,cm^2$ (d) $3\frac{1}{2}\,cm^2$
(e) $4\frac{1}{2}\,cm^2$

A3 C, A, B

A4 (a) $16\,cm^2$ to $18\,cm^2$ (b) $29\,cm^2$ to $33\,cm^2$
(c) $53\,cm^2$ to $61\,cm^2$

A5 (a) 6 cm (b) 8 cm (c) 12 cm (d) 10 cm

A6 (a) There are many possibilities.
(b)

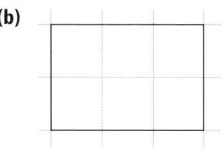

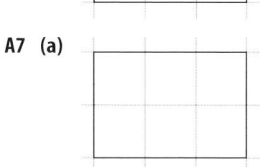

A7 (a)

(b) There are many possibilities.

A8 (a)

(b) There are many possibilities.
(c) There are many possibilities.

B Area of rectangle and right-angled triangle (p 100)

B1 (a) 3 (b) 7 (c) $21\,cm^2$

B2 (a) $16\,cm^2$ (b) $14\,cm^2$ (c) $10\,cm^2$

B3 (a) 4 × 3 (b) $12\,cm^2$

B4 (a) $18\,cm^2$ (b) $15\,cm^2$ (c) $12\,cm^2$ (d) $4\,cm^2$

B5 (a) $49\,cm^2$ (b) $100\,cm^2$ (c) $225\,cm^2$ (d) $300\,cm^2$
(e) $132\,cm^2$

B6 3 cm

B7 1 cm by 72 cm, 2 cm by 36 cm, 3 cm by 24 cm,
4 cm by 18 cm, 6 cm by 12 cm, 8 cm by 9 cm

B8 (a) $18\,cm^2$ (b) $9\,cm^2$

B9 (a) (i) $6\,cm^2$ (ii) $3\,cm^2$
(b) (i) $9\,cm^2$ (ii) $4\frac{1}{2}\,cm^2$
(c) (i) $20\,cm^2$ (ii) $10\,cm^2$

C Area of a shape made from simpler shapes (p 102)

C1 (a) 8 cm² (b) 4 cm² (c) 12 cm²
C2 (a) 3 cm² (b) 12 cm² (c) 15 cm²
C3 (a) 16 cm² (b) 14 cm²
C4 (a) A: 6 cm², B: 6 cm², L-shape: 12 cm²
(b) C: 10 cm², D: 2 cm², L-shape: 12 cm²
(c) 16 cm
C5 (a) Sketches showing two ways of splitting into rectangles
(b) 15 cm²
(c) 18 cm
C6 (a) (i) A sketch (ii) 17 cm² (iii) 20 cm
(b) (i) A sketch (ii) 14 cm² (iii) 16 cm
C7 Perimeter = 28 m, area = 42 m²

D Using decimals (p 104)

D1 (a) 6 m² (b) 3 m² (c) 10.5 m²
D2 (a) A sketch (b) 7 m² and 3 m² (c) 10 m²
D3 A calculator check
D4 (a) 5.7 m² (b) 9.9 m² (c) 12.9 m² (d) 4.4 m²
D5 1.17 m²
D6 2.99 m²
D7 (a) (i) 14 cm² (ii) 7 cm²
(b) (i) 15 cm² (ii) 7.5 cm²
D8 (a) 4.5 cm² (b) 6.6 cm²

Test yourself (p 106)

T1 (a) 8 cm²
(b) (i) 8 cm² (ii) 18 cm
(c) 11 cm² or 12 cm²
T2 39 cm²
T3 32 cm
T4 (a) 2 cm (b) 7 cm (c) 8 cm
T5 (a) 22 cm² (b) 14.4 cm² (c) 9 cm² (d) 5 cm²

24 Probability

A The probability scale (p 108)

A1 (a) Blue
(b) Black
(c) A black, B yellow, C blue
A2 (a) More than $\frac{1}{2}$
(b) Scale from 0 to 1 with $\frac{1}{2}$ marked; arrows: blue, yellow, white
A3 (a) Black (b) Blue (c) 0
A4 (a) D (b) E (c) B (d) C (e) A

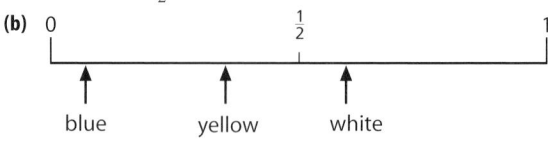

B Equally likely outcomes (p 110)

B1 (a) (i) $\frac{1}{6}$ (ii) $\frac{2}{6} = \frac{1}{3}$ (iii) $\frac{3}{6} = \frac{1}{2}$
(b) Scale from 0 to 1 with $\frac{1}{2}$ marked; arrows: black, white, blue
(c) (i) 4 (ii) $\frac{4}{6} = \frac{2}{3}$
B2 (a) $\frac{1}{5}$ (b) $\frac{3}{5}$ (c) $\frac{2}{5}$ (d) $\frac{2}{5}$ (e) 1
B3 (a) $\frac{2}{12} = \frac{1}{6}$ (b) $\frac{1}{12}$ (c) $\frac{5}{12}$ (d) $\frac{3}{12} = \frac{1}{4}$
(e) $\frac{2}{12} = \frac{1}{6}$ (f) $\frac{7}{12}$ (g) $\frac{5}{12}$
B4 (a) $\frac{5}{15} = \frac{1}{3}$ (b) $\frac{4}{15}$ (c) $\frac{6}{15} = \frac{2}{5}$
(d) $\frac{10}{15} = \frac{2}{3}$ (e) $\frac{9}{15} = \frac{3}{5}$
B5 (a) $\frac{12}{20} = \frac{3}{5}$ (b) $\frac{8}{20} = \frac{2}{5}$ (c) $\frac{11}{20}$ (d) $\frac{9}{20}$
(e) $\frac{6}{20} = \frac{3}{10}$ (f) $\frac{5}{20} = \frac{1}{4}$ (g) $\frac{7}{20}$ (h) $\frac{2}{20} = \frac{1}{10}$
B6 (a) $\frac{1}{8}$ (b) $\frac{2}{8} = \frac{1}{4}$ (c) $\frac{3}{8}$
(d) $\frac{4}{8} = \frac{1}{2}$ (e) $\frac{5}{8}$
B7 The game is not fair because Mark, Sam and Danielle do not have the same chance of winning. Mark is most likely to win $\left(\frac{4}{8}\right)$, then Sam $\left(\frac{3}{8}\right)$, then Danielle $\left(\frac{1}{8}\right)$.

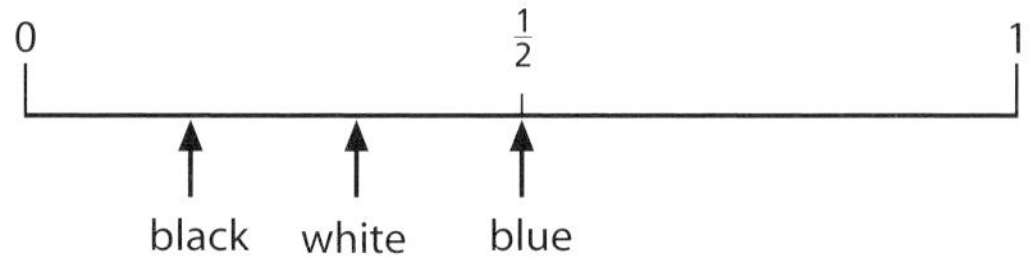

Test yourself (p 112)

T1 Scale from 0 to 1 with arrows: B, A, C
T2 (a) X (b) V (c) W (d) Z (e) Y
T3 (a) $\frac{1}{8}$ (b) $\frac{2}{8} = \frac{1}{4}$ (c) $\frac{3}{8}$ (d) $\frac{6}{8} = \frac{3}{4}$ (e) 0

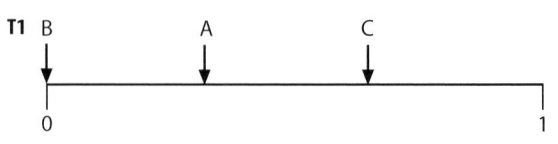

Review 4 (p 113)

1 (a) (i) 32
   (ii) An explanation such as:
   'You add 5 to get from one number to the next and 27 + 5 = 32. So 32 is the next number.'
   (b) 100 is not in the pattern with an explanation such as:
   'All the numbers in the pattern end in a 2 or a 7 and 100 ends in a 0 so 100 cannot be a number in the pattern.'

2 17 metres

3 (a) 175 °C   (b) 42 °C   (c) 22.4 °C

4 1.5 kg

5 (a) 15 cm²   (b) $10\frac{1}{2}$ cm² or 10.5 cm²
   (c) $37\frac{1}{2}$ m² or 37.5 m²

6 $\frac{1}{6}$

7 (a) 5.7   (b) 4.2   (c) 3.7   (d) 2.4

8 1.8

9 (a) $\frac{1}{4}$ (b) $\frac{1}{2}$ or $\frac{2}{4}$

10 31, 43

25 Enlargement

A Enlargement on squared paper (p 114)

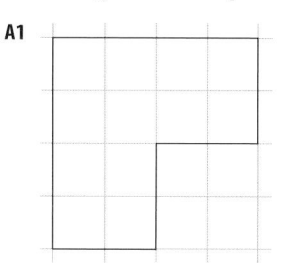

A1

A2 The required enlargements

A3 (a) You multiply its length by 3.
   (b) The required enlargements

B Scale factor (p 115)

B1 C (scale factor 2) and E (scale factor 3) are correct.

B2 (a) Base of P = 2 cm
       Base of Q = 6 cm
       So a scale factor of 3 has been used.
   (b) Height of P = $1\frac{1}{2}$ cm
       Height of Q = $4\frac{1}{2}$ cm
       This fits with a scale factor of 3.
   (c) 4

B3 (a) 16 cm
   (b) The required drawing; perimeter = 32 cm
   (c) The required drawing; perimeter = 48 cm
   (d) It is multiplied by the scale factor.

B4 (a) 2 cm   (b) 1 cm   (c) 6 cm

B5 3 cm

B6 4

Test yourself (p 116)

T1 The required drawing

T2 P: 2, Q: 3

26 Negative numbers

A Putting temperatures in order (p 117)

A1 (a) P: 4°C, Q: ⁻4°C, R: 8°C, S: ⁻6°C
(b) S
(c) R
(d) ⁻6°C, ⁻4°C, 4°C, 8°C

A2 (a) Thursday (b) Friday (c) Wednesday
(d) ⁻5°C, ⁻3°C, ⁻2°C, 1°C, 3°C

A3 (a) ⁻4°C, ⁻2°C, 0°C, 2°C, 3°C
(b) ⁻10°C, ⁻6°C, ⁻2°C, 3°C, 5°C

A4 Saturday

A5 (a) 6°C (b) ⁻2°C
(c) Wednesday (d) Tuesday and Friday
(e) Tuesday and Wednesday (f) Friday

A6 (a) Edinburgh (b) London (c) Glasgow
(d) ⁻4°C, ⁻3°C, ⁻2°C, 3°C, 4°C, 5°C

B Temperature changes (p 119)

B1 (a) 2°C (b) ⁻6°C
(c) Temperature **A** is warmer by **8** degrees.

B2 4 degrees

B3 2°C

B4 ⁻1°C

B5 (a) 08:18 (b) 00:18
(c) 2 degrees (d) 1 degree

B6 (a) 90 degrees (b) 540 degrees
(c) Jupiter (d) ⁻230°C

C Negative coordinates (p 120)

C1 (a) (⁻3, 1) (⁻2, 2) (2, ⁻1) (3, 0) (1, 0)
(b) (4, ⁻1) (⁻2, ⁻1) (⁻3, 3) (2, ⁻1) (2, ⁻1) (3, 0) (3, ⁻2)
(c) (⁻2, ⁻1) (⁻1, ⁻2) (2, 2) (2, 2) (⁻1, 1) (1, 0)
(d) (⁻1, 0) (⁻3, 3) (⁻3, 0) (⁻3, 0) (⁻3, 3) (2, ⁻1) (3, 3)

C2 (a) Bear (b) Lion (c) Tiger
(d) Badger (e) Wolf (f) Jaguar

Test yourself (p 121)

T1 (a) (i) 7°C (ii) ⁻10°C
(b) (i) 6 degrees (ii) 8 degrees
(c) ⁻7°C

T2 (a) (i) ⁻7°C (ii) 14°C
(b) ⁻11°C

T3 (a) (i) I (ii) C
(b) (i) (1, ⁻2) (ii) (0, 1)

27 Mean

A Finding the mean of a data set (p 122)

A1 (a) 392 kg (b) 56 kg
A2 (a) 116 kg (b) 14.5 kg
A3 65 cm
A4 75.4p
A5 13°C
A6 (a) 37 (b) 76 (c) 3.925
(d) 27.1 (e) 280.5
A7 (a) 15 (b) 15.3
A8 (a) 2.94 km (b) 2.66 km (c) 2.44 km
(d) 3.3 km (e) 3.35 km

B Comparing two sets of data (p 124)

B1 (a) £17 (b) £17.75 (c) Girls
B2 (a) £10 (b) £15 (c) Girls
B3 (a) Rovers 2.5, United 2.2 (b) Rovers
B4 (a) 15.75 (b) 15 (c) Last term
B5 (a) (i) 144 cm (ii) 17 cm
(b) (i) 151 cm (ii) 21 cm
(c) Boys
(d) Boys

B6 (a) B. The mean for B, £26.3k, is greater than the mean for A, £26k.
(b) B. The range for B, £28k, is greater than the range for A, £25k.

B7 (a) 3.25 eggs (b) 3.5 eggs

Test yourself (p 125)

T1 (a) 43 (b) 33.6
T2 (a) (i) 58 g (ii) 55.5 g
(b) (i) 50 g (ii) 53 g
(c) First plant
(d) Second plant

28 Starting equations

A Arrow diagrams (p 126)

A1 (a) 11 (b) 7 (c) 12 (d) 3
A2 (a) 4 (b) 5 (c) 8 (d) 15
A3 (a) 9 (b) 3

B Think of a number (p 127)

B1 4
B2 3
B3 (a) 2 (b) 3 (c) 4 (d) 7 (e) 15 (f) 14
B4 (a) 9 (b) 25 (c) 40

C Number puzzles (p 128)

C1 (a) 6 (b) 7 (c) 2 (d) 10 (e) 5
(f) 10 (g) 8 (h) 20 (i) 5
C2 (a) 9 (b) 2 (c) 2 (d) 4 (e) 3 (f) 10
C3 (a) 3 (b) 5 (c) 11 (d) 7 (e) 2 (f) 24
C4 (a) 4 (b) 2 (c) 30

D Using letters (p 128)

D1 (a) 4 (b) 6 (c) 2 (d) 6
D2 (a) 6 (b) 2 (c) 12 (d) 8
D3 (a) 8 (b) 7 (c) 6 (d) 1
D4 (a) 18 (b) 3 (c) 16 (d) 10
D5 (a) $n = 2$ (b) $n = 1$ (c) $n = 13$ (d) $n = 7$
(e) $n = 3$ (f) $n = 1$ (g) $n = 35$ (h) $n = 55$
D6 (a) $n = 10$ (b) $x = 12$ (c) $y = 4$ (d) $k = 5$
D7 (a) $n = 5$ (b) $m = 7$ (c) $p = 8$ (d) $q = 1$
D8 (a) $n = 5$ (b) $n = 9$ (c) $n = 28$ (d) $n = 50$
(e) $x = 7$ (f) $x = 7$ (g) $x = 7$ (h) $x = 5$
(i) $a = 22$ (j) $a = 8$ (k) $a = 7$ (l) $a = 20$

Test yourself (p 129)

T1 (a) 16 (b) 6 (c) 8 (d) 5
T2 (a) $a = 5$ (b) $b = 2$ (c) $c = 4$
T3 (a) $x = 20$ (b) $x = 6$ (c) $x = 8$

29 Finding your way

A Using a town plan (p 130)

A1 (a) Corporation Street
(b) Boulevard Street
(c) (i) A3 (ii) C2 (iii) A1 (iv) C3
A2 Left
A3 (a) Left (b) Right (c) Left
A4 (a) East (b) North-east (c) South-east
A5 (a) South (b) North-west (c) South-west
A6 The railway station
A7 Alexandra Road, Park Avenue, Mill Street, Grays Inn Road
A8 (a) Directions from the post office to the station
(b) Directions from the cinema to the hospital
(c) Directions from the railway station to the fire station
A9 (a) 450 m (b) 850 m

Test yourself (p 132)

T1 (a) C2 (b) A1
T2 Right
T3 (a) East (b) South-west
T4 In the market square, turning anticlockwise
T5 The library
T6 Directions from the library to the school
T7 500 m

Review 5 (p 133)

1. **(a)** The edges measure 3 cm, 4 cm and 5 cm.
   The perimeter is 12 cm.
   (b) Scalene
   (c) **(i)** An accurate drawing of a right-angled triangle with edges of 6 cm, 8 cm and 10 cm
   (ii) 24 cm
   (d) 24 cm^2
2. $-2\,°C$
3. 2
4. $-4\,°C, -2\,°C, 0\,°C, 1\,°C, 3\,°C$
5. **(a)** Ice cream stall **(b)** Duck pond
   (c) North-west
6. **(a)** $n = 3$ **(b)** $x = 9$ **(c)** $m = 9$ **(d)** $y = 15$
7. **(a)** Alex: 3.6 cups, Becky: 4 cups
   (b) Alex: 3 cups, Becky: 6 cups

30 Volume

A Counting cubes (p 134)

A1 **(a)** 3 cm^3 **(b)** 5 cm^3 **(c)** 4 cm^3 **(d)** 6 cm^3
A2 4
A3 **(a)** $2 \times 3 = 6$ cm^3 **(b)** $3 \times 6 = 18$ cm^3
A4 **(a)** 12 cm^3 **(b)** 20 cm^3
A5 **(a)** Starting from the top: 1, 4, 9, 16
   (b) 30 cm^3
A6 **(a)** 8 **(b)** 3 **(c)** $3 \times 8 = 24$ cm^3

Test yourself (p 135)

T1 **(a)** 6 cm^3 **(b)** 15 cm^3 **(c)** 24 cm^3

31 Evaluating expressions

A Simple substitution (p 136)

A1 **(a)** 6 **(b)** 9 **(c)** 1 **(d)** 0
A2 **(a)** 6 **(b)** 15 **(c)** 12 **(d)** 30
A3 **(a)** 6 **(b)** 4 **(c)** 3 **(d)** 2
A4 **(a)** 6 **(b)** 30 **(c)** 20 **(d)** 3
A5

	$n = 2$	$n = 4$	$n = 6$	$n = 8$
$n + 3$	5	7	9	11
$n - 2$	0	2	4	6
$3n$	6	12	18	24
$n \div 2$	1	2	3	4

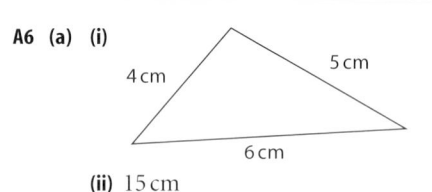

A6 **(a)** **(i)** [triangle with sides 4 cm, 5 cm, 6 cm]
   (ii) 15 cm
   (b) 23 cm

B Rules for calculation (p 137)

B1 **(a)** 18 **(b)** 21 **(c)** 12 **(d)** 7 **(e)** 17 **(f)** 19
   (g) 30 **(h)** 5 **(i)** 2
B2 **(a)** 10 **(b)** 2 **(c)** 12 **(d)** 2 **(e)** 2 **(f)** 14
B3 **(a)** 28 **(b)** 20 **(c)** 6 **(d)** 5 **(e)** 29 **(f)** 21
B4 **(a)** 12 **(b)** 4 **(c)** 2 **(d)** 5 **(e)** 7 **(f)** 7
B5 **(a)** 2 **(b)** 3 **(c)** 7
B6 **(a)** 10 **(b)** 3 **(c)** 10 **(d)** 2 **(e)** 15 **(f)** 2

C Substituting into linear expressions (p 138)

C1 **(a)** 9 **(b)** 20 **(c)** 6 **(d)** 18
C2 **(a)** 28 **(b)** 39 **(c)** 30 **(d)** 15
C3 **(a)** 4 **(b)** 4 **(c)** 2 **(d)** 2
C4 **(a)** 1 **(b)** 2 **(c)** 2 **(d)** 9
C5 **(a)**

	$n = 3$	$n = 6$	$n = 9$	$n = 12$
$3n + 4$	13	22	31	40
$5n - 2$	13	28	43	58
$2(n - 1)$	4	10	16	22
$4(1 + n)$	16	28	40	52
$\frac{n + 6}{3}$	3	4	5	6
$\frac{n}{3} - 1$	0	1	2	3

(b) (i) $4(1 + n)$ (ii) $3n + 4$ and $5n - 2$
(iii) $5n - 2$ and $4(1 + n)$ (iv) $5n - 2$
(v) $\frac{n}{3} - 1$

C6 (a) (i)

```
        5 cm
  ┌──────────────┐
2 cm             │
  └──────────────┘
```

(ii) $10\,\text{cm}^2$

(b) $290\,\text{cm}^2$

D Formulas in words (p 140)

D1 (a) 150 cm (b) 162 cm
D2 (a) 200 mm (b) 480 mm (c) 760 mm
   (d) 1180 mm
D3 (a) 240 mm (b) 340 mm (c) 440 mm
   (d) 640 mm
D4 (a) 8 miles (b) 3 miles
D5 (a) 6 (b) 9 (c) 16
D6 (a) £120 (b) £210 (c) £270
D7 91
D8 (a) £54 (b) £168
D9 (a) £22.40 (b) £80.88

E Formulas without words (p 142)

E1 (a) 10 (b) 14
E2 (a) 10 cm (b) 35 cm (c) 40 mm
E3 (a) 480 grams (b) 680 grams (c) 1080 grams
E4 (a) £35 (b) £80

Test yourself (p 143)

T1 (a) 12 (b) 7 (c) 17
T2 (a) 2 (b) 25
T3 (a) 6 (b) 45 (c) 14 (d) 2
T4 (a) 7 (b) 10 (c) 11 (d) 6
T5 (a) £70 (b) £66
T6 (a) 20 (b) 35 (c) 53

32 Estimating and calculating with money

A Solving problems without a calculator (p 144)

A1 (a) £13 (b) £12.50 (c) £13 (d) £32
A2 £3.50
A3 £5
A4 It is £1.50 cheaper to buy a family ticket.
A5 (a) £2.50 (b) £1.25 (c) 3
A6 (a) 3 (b) 30p or £0.30
A7 (a) 2 (b) 4

B Estimating answers (p 145)

B1 (a) £36 (b) 2
B2 (a) 7 (b) £2.30 (c) Yes
B3 (a) £12
   (b) Less, because a child ticket is just under £5, so 10 tickets would be just under £50.
   (c) More, because an adult ticket is roughly £6 and a child ticket is roughly £5, so tickets for two adults and two children would be roughly £22.

C Using a calculator (p 146)

C1 (a) £17.50 (b) £22.80
C2 (a) 3 (b) £8.37
C3 £29
C4 (a) £23.94 (b) £6.06
C5 3
C6 £1.44
C7 (a) £180 (b) £135

Test yourself (p 147)

T1 £12
T2 3
T3 £6.60
T4 (a) 4 (b) £4.20

33 Capacity

A Litres, millilitres and other metric units (p 148)

A1 250 ml

A2 500 ml

A3 1.7 litres

A4 20 litres

A5 (a) 2 litres (b) 1.5 litres (c) 50 litres

A6 0.6 litres or 600 ml

A7 750 ml

A8 (a) 0.75 litre (b) 0.5 litre
      (c) 0.8 litre (d) 0.1 litre

A9 (a) 1500 ml (b) 5200 ml
      (c) 10 000 ml (d) 400 ml

A10 (a) 3 litres (b) 4.5 litres
       (c) 1.65 litres (d) 6.2 litres

A11 100

A12 50 ml, $\frac{1}{2}$ litre, 750 ml, 1.2 litres, 1500 ml

A13 (a) m (b) ml (c) cm (d) kg

Test yourself (p 149)

T1 4 litres, 1.5 litres, 500 ml, 250 ml

T2 0.4 litre or 400 millilitres

T3 (a) (i) 8 litres (ii) 3.5 litres
      (b) 1250 ml

T4 (a) g (b) mm

34 Drawing and using graphs

A Tables and graphs (p 150)

A1 (a) 9 cm

(b)
Time (min)	0	1	2	3	4	5
Depth (cm)	5	7	**9**	**11**	**13**	**15**

(c), (d)

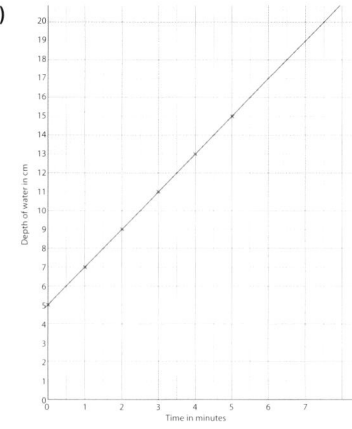

(e) 19 cm

(f) 12 cm

(g) $2\frac{1}{2}$ minutes

(h) 21 minutes past 3

A2 (a) 18 cm

(b)
Time in hours	0	1	2	3	4
Height of candle in cm	26	22	**18**	**14**	**10**

(c), (d)

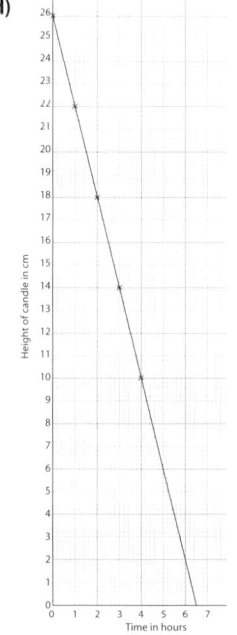

(e) 2 cm

(f) 8 cm

- (g) $3\frac{1}{2}$ hours
- (h) $6\frac{1}{2}$ hours
- (i) 8 o'clock

A3 (a)

Time (min)	0	5	10	15	20	25
Temp. (°C)	20	28	**36**	**44**	**52**	**60**

(b), (c)

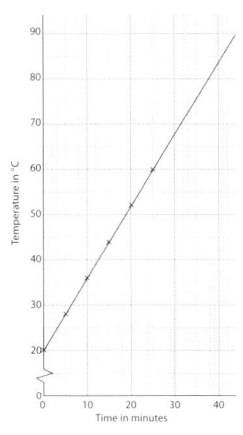

- (d) 84 °C
- (e) 6 or 7 minutes
- (f) At about 4:46

A4 (a) 11 °C

(b) It will have dropped by 20 degrees, and its temperature will be ⁻5 °C.

(c)

Time (h)	0	1	2	3	4	5
Temp. (°C)	15	**11**	**7**	**3**	**⁻1**	**⁻5**

(d)

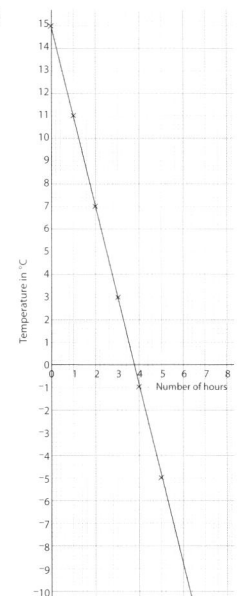

- (e) 5 °C
- (f) ⁻9 °C
- (g) $4\frac{1}{2}$ hours
- (h) 3:30 p.m.

B Graphs and rules (p 154)

B1 (a) Number of kilometres
= 10 × 8 ÷ 5
= 80 ÷ 5 = 16

(b) (i) 32 km   (ii) 48 km   (iii) 80 km

(c)

Miles	0	10	20	30	40	50
km	0	16	**32**	**48**	**64**	**80**

(d)

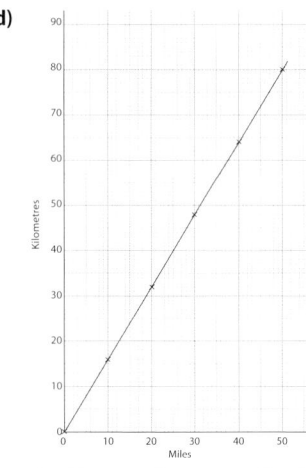

(e) (i) About 67 km   (ii) About 88 km

(f) (i) 12 or 13 miles   (ii) About 47 miles

(g) (Answers are approximate.)

Dunkerque	17 or 18
Boulogne	21
St Omer	28
Centre ville	5
Paris	53
Versailles	49
Aéroport d'Orly	41
Fontainebleu	22 or 23
Gare TGV	9

B2 (a)

m	1	2	3	4	5	6
c	**5**	**9**	**13**	**17**	**21**	**25**

Answers: Chapter 34 213

(b) [graph: Z car taxis, Charge in £ (c) vs Distance travelled in miles (m)]

(c) (Answers are approximate.)
- **(i)** £8.20
- **(ii)** £23.00
- **(iii)** £3.80
- **(iv)** 3.5 miles
- **(v)** 5.6 miles
- **(vi)** 3.2 miles

B3 (a)

m	1	2	3	4	5	6
c	7.50	10.50	13.50	16.50	19.50	22.50

(b) [graph: Z car taxis and Aardvark taxis, Charge in £ (c) vs Distance travelled in miles (m)]

(c) **(i)** £7.00   **(ii)** £9.00

(d) Z cars are cheaper.

(e) Aardvark are cheaper.

(f) 3.5 miles

(g) Phone **Aardvark** for journeys over **3.5** miles as they are cheaper.

Test yourself (p 157)

T1 (a)

Time (min)	1	2	3	4	5	6
Charge (£)	**8**	11	**14**	17	**20**	23

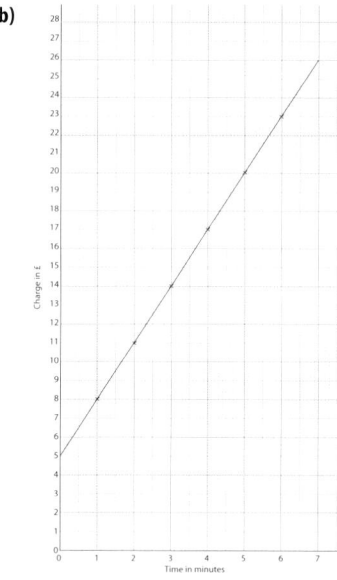

(b) [graph: Charge in £ vs Time in minutes]

(c) £15.50

(d) $5\frac{1}{2}$ minutes

(e) **(i)** $\frac{1}{2}$ minute or 30 seconds   **(ii)** $4\frac{1}{2}$ minutes
(iii) £24.50

Review 6 (p158)

1 $10\,\text{cm}^3$

2 (a) 7   (b) 23   (c) 21   (d) 2   (e) 1

3 (a) 3   (b) £42

4 2800 ml

5 (a) £8.00

(b)

Weight in kg (w)	2	4	6	8	10	12
Delivery cost in £ (c)	4.00	**5.00**	**6.00**	**7.00**	**8.00**	**9.00**

(c)

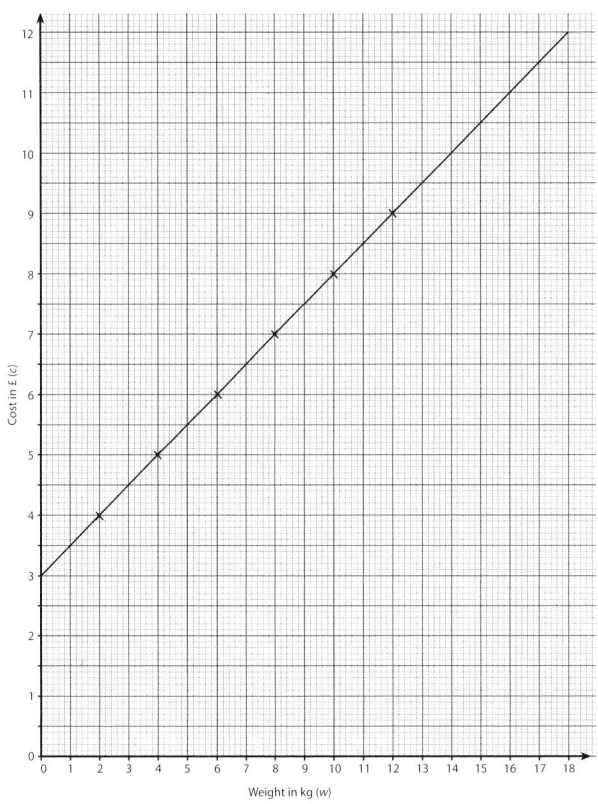

(d) £4.70

(e) £10.50

(f) Any answer between 8.4 kg and 8.6 kg is acceptable.

6 (a) grams   (b) litres

7 £3.50

35 Fractions, decimals and percentages

A Fraction and percentage equivalents (p 159)

A1 (a) $\frac{1}{2}$ is equivalent to **50%**.

(b) $\frac{1}{4}$ is equivalent to **25%**.

(c) $\frac{1}{10}$ is equivalent to 10%.

(d) $\frac{3}{4}$ is equivalent to 75%.

A2 (a) (i) $\frac{3}{4}$ (ii) 75%

(b) (i) $\frac{1}{2}$ (ii) 50%

(c) (i) $\frac{1}{4}$ (ii) 25%

A3 $\frac{1}{4}$ is bigger; it is equivalent to 25%.

A4 (a) $\frac{1}{10}$, 50%, $\frac{3}{4}$, 100%   (b) 10%, $\frac{1}{4}$, $\frac{1}{2}$, 75%

B Fraction and decimal equivalents (p 161)

B1 (a) $\frac{2}{10}$ (b) $\frac{9}{10}$ (c) $\frac{3}{10}$ (d) $\frac{7}{10}$

B2 0.1 and $\frac{1}{10}$, $\frac{8}{10}$ and 0.8, 0.5 and $\frac{5}{10}$, $\frac{6}{10}$ and 0.6

B3 (a) 0.1, $\frac{3}{10}$, 0.8, $\frac{9}{10}$ (b) $\frac{2}{10}$, 0.4, 0.6, $\frac{7}{10}$

B4 (a) Any decimal bigger than 0.25 and smaller than 0.5

(b) Any decimal bigger than 0.5 and smaller than 0.75

(c) Any decimal smaller than 0.25

B5 (a) 0.5 kg   (b) $\frac{1}{4}$ litre

B6 (a) $\frac{1}{2}$, 0.6, 0.75, $\frac{8}{10}$ (b) $\frac{1}{4}$, 0.3, 0.4, $\frac{1}{2}$

C Fraction, decimal and percentage equivalents (p 162)

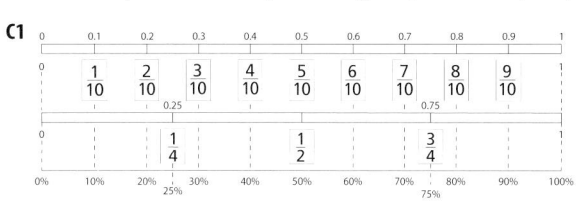

C1

C2

Fraction		Decimal		Percentage
$\frac{1}{2}$	=	**0.5**	=	50%
$\frac{1}{4}$	=	0.25	=	**25%**
$\frac{1}{10}$	=	**0.1**	=	**10%**
$\frac{2}{10}$	=	**0.2**	=	20%

C3 (a) $\frac{3}{4}$, 0.75, 75%; $\frac{4}{10}$, 0.4, 40%; $\frac{3}{10}$, 0.3, 30%

(b) 0.7 is the odd one out; equivalents are $\frac{7}{10}$ and 70%.

C4 (a) $\frac{6}{10}$ (b) 0.6   (c) 60%   (d) 40%

C5 (a) (or equivalent)

(b) $\frac{7}{10}$
  (c) 70%
- **C6** (a) BLUE (b) GREY (c) GREEN (d) PURPLE
- **C7** 0.6 is bigger; it is equivalent to 60%.
- **C8** (a) 20%, $\frac{1}{4}$, 0.4, 60% (b) $\frac{3}{10}$, 0.5, 80%, $\frac{9}{10}$

D Percentage of a quantity, mentally (p 163)

- **D1** (a) 4 (b) 6 (c) 20 (d) 50 (e) 100
- **D2** (a) 2 (b) 3 (c) 10 (d) 25 (e) 50
- **D3** (a) 6 (b) 9 (c) 30 (d) 75 (e) 150
- **D4** (a) 5 (b) 3 (c) 20 (d) 25 (e) 14
- **D5** (a) 10 (b) 6 (c) 40 (d) 50 (e) 28
- **D6** 10% of 40 = 20% of 20
  20% of 50 = 25% of 40
  50% of 6 = 10% of 30
  25% of 60 = 50% of 30
- **D7** (a) 18 (b) 14 (c) 6 (d) 24 (e) 24
- **D8** (a) 4 (b) 2 (c) 6 (d) 12 (e) 14
- **D9** 250 g
- **D10** 40
- **D11** 45
- **D12** (a) 60 (b) 30 (c) 24 (d) 6

Test yourself (p 164)

- **T1** (a) $\frac{3}{10}$
  (b) 30%
  (c) (or equivalent)
- **T2** (a) $\frac{1}{2}$ or $\frac{1}{5}$ (b) $\frac{6}{10}$ or $\frac{3}{5}$ (c) 0.75 (d) 70%
- **T3** (a) 6 (b) 18 (c) 8 (d) 4
- **T4** One more girl wears glasses.

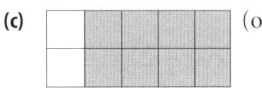

36 Two-way tables

A Reading tables (p 165)

- **A1** Emily
- **A2** Jessica
- **A3** (a) Third (b) 2005
- **A4** Olivia
- **A5** 39%
- **A6** 31%
- **A7** (a) True (b) False (c) False (d) True
- **A8** Countdown
- **A9** (a) 30 minutes (b) 45 minutes
  (c) 1 hour (d) 25 minutes
  (e) 2 hours (f) 2 hours 10 minutes
- **A10** BBC1: CBBC BBC2: Flog It!
  ITV1: CITV C4: Countdown

B Distance tables (p 166)

- **B1**

Cork							
183	Dublin						
126	143	Galway					
62	127	64	Limerick				
188	80	63	126	Longford			
110	79	112	48	78	Roscrea		
75	190	127	63	189	111	Tralee	
71	112	145	81	192	129	144	Waterford

- **B2** (a) 62 miles (b) 152 miles (c) 214 miles
- **B3** (a) 120 miles (b) Aberdeen and Gatwick
- **B4** (a) 445 miles (b) 460 miles (c) 15 miles

Test yourself (p 167)

- **T1** (a) 302 miles (b) 164 miles (c) 668 miles

37 Scale drawings

A Simple scales (p 168)

A1 (a)

Side	Length on blue shape	Length on scale drawing
AB	8 cm	4 cm
BC	**4 cm**	**2 cm**
CD	**4 cm**	**2 cm**
DE	**2 cm**	**1 cm**
EF	**4 cm**	**2 cm**
FA	**6 cm**	**3 cm**

    (b) 2 cm   **(c)** 5 cm   **(d)** 10 cm

A2 (a)

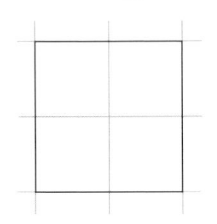

    (b) 5.6 cm   **(c)** 2.8 cm

A3 (a) 5 cm   **(b)** 12 cm   **(c)** 6.5 cm   **(d)** 13 cm

A4 (a) Scale drawing on centimetre squared paper
    (b) More   **(c)** 7.6 cm   **(d)** 15.2 cm

A5 (a) Scale drawing on centimetre squared paper
    (b) 12.5 cm   **(c)** 25 cm

A6 (a) 1 cm   **(b)** 5 cm
    (c) 1 cm represents **5** cm.   **(d)** 2 cm
    (e) 10 cm

A7 (a)

    (b) 5 cm   **(c)** 25 cm   **(d)** 53°

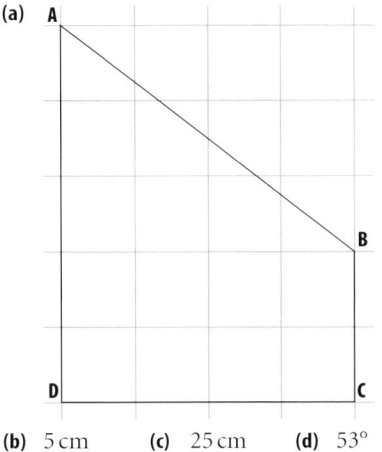

A8 (a), (b) Scale drawing with the tap 3 cm from A
    (c) 17 cm   **(d)** 85 m

A9 (a) Scale drawing on centimetre squared paper
    (b) 5.8 cm   **(c)** 58 cm   **(d)** 6.4 cm   **(e)** 64 cm

A10 (a) 14.3 cm   **(b)** 143 km   **(c)** 67 km

B Harder scales (p 171)

B1 (a) 8 cm   **(b)** 4 m   **(c)** 2 cm   **(d)** 1 m
    (e) 4 cm   **(f)** 2 m   **(g)** 2 cm by 3 cm

B2 (a) 10 cm   **(b)** 2 m   **(c)** 20 cm   **(d)** 1 cm
    (e) 20 cm   **(f)** 80 cm

Test yourself (p 172)

T1 (a) 35 cm   **(b)** 5 cm   **(c)** 25 cm   **(d)** 3 cm

T2 (a) 40 cm   **(b)** 4.5 cm   **(c)** 45 cm

T3 (a) Scale drawing   **(b)** 3.7 m (to the nearest 0.1 m)

38 Using percentages

A Percentage bars (p 173)

A1 (a) 37% (b) 63% (c) 28% (d) 35%

A2

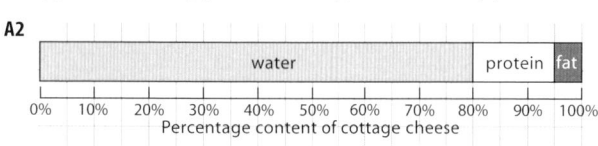

| 0% | 10% | 20% | 30% | 40% | 50% | 60% | 70% | 80% | 90% | 100% |

water | protein | fat

Percentage content of cottage cheese

B Interpreting pie charts (p 174)

B1 (a) 76% (b) 13%
(c) 11% (d) A: true B: true C: false
(e) About 45 g (f) About 6 g or 7 g

B2 (a) Clothes (b) 30%
(c) 27% (d) About £10 or £11

B3 (a) 28% (b) 13%
(c) A: true B: false C: true

B4 (a) Housing and fuel; 25%
(b) Transport
(c) 22%

C Drawing pie charts (p 176)

C1 pie chart with water, carbohydrate, fat, protein

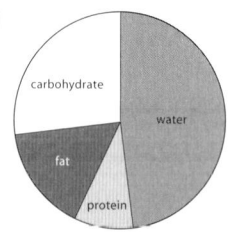

C2 pie chart with theft, burglary, violence, criminal damage, fraud, other

C3 pie chart with European Union, other Europe, North America, others

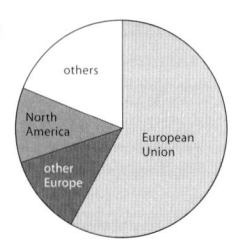

D Comparing (p 177)

D1 (a) 20% (b) 34%

D2 Clothing; magazines, books and stationery; mobile phones; other

D3

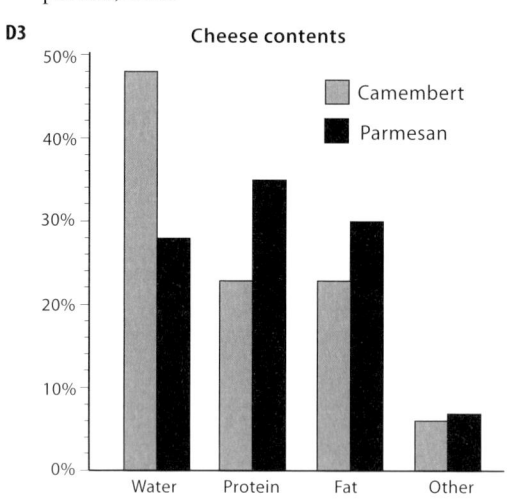

Cheese contents bar chart (Camembert, Parmesan) — Water, Protein, Fat, Other

Camembert has a much higher water content.
Parmesan has higher protein and fat content.

Test yourself (p 178)

T1 (a) pie chart with rock, pop, urban, MOR, dance, other

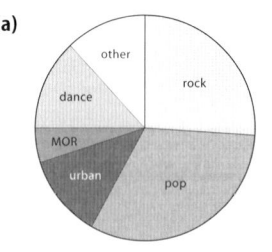

(b) Comments such as:
- Pop was the most popular in 2000, but rock was most popular in 2005.
- Urban had about the same proportion of sales in both years.
- The proportion of sales of MOR almost doubled from 2000 to 2005.

T2 (a) Saturday
(b) Features such as:
- The women generally shop more than men.
- The women do more of their shopping on weekdays compared with men.
- The men's weekday shopping starts very low on Monday and builds up gradually through the week, but the women's weekday shopping changes little from day to day.

39 Conversion graphs

Answers for this chapter will be approximate.

A Using a conversion graph (p 179)

A1 (a) 32 km (b) 43 km (c) 13 km (d) 28 km

A2 22 miles

A3 (a) 25 miles (b) 15 miles (c) 6 miles

A4 You could convert 25 miles and then double (result 80 km).

A5 (a) 160 km
 (b) 120 km
 (c) 37 or 38 miles (37.5 miles)
 (d) 62 or 63 miles (62.5 miles)

A6 88 km

A7 (a) (i) 2.5 or 2.6 litres (ii) 1.6 litres
  (iii) 1 litre
 (b) (i) 5.3 pints (ii) 2.6 pints
  (iii) 1.6 pints
 (c) About 8.5 litres

A8 (a) (i) 1.8 kg (ii) 2.5 kg
  (iii) 5.2 or 5.3 pounds (iv) 2 pounds
 (b) (i) 5.5 kg (ii) 22 pounds

B Drawing a conversion graph (p 181)

B1 (a) A graph on grid A of sheet FT–26
 (b) (i) 31 km/h (ii) 21 or 22 knots

B2 (a) A graph on grid B
 (b) (i) About 96 km^2 (ii) About 33 square miles

B3 (a) A graph on grid C
 (b) (i) 6.2 m^3 (ii) About 190 cubic feet

B4 (a) A graph on grid D
 (b) (i) €28 (ii) About £22
  (iii) About £12

B5 (a) A graph on grid A of sheet FT–27
 (b) 27 °C
 (b) 14 °F

Test yourself (p 182)

T1 (a) $8 (b) £10.50

T2 (a) A graph on grid B of sheet FT–27
 (b) (i) 1.7, 1.75 or 1.8 gallons
  (ii) 13.6 litres
 (c) 2.2, 2.25 or 2.3 litres

Review 7 (p 183)

1 (a) 0.3 (b) $\frac{3}{4}$ (c) $\frac{4}{10}$ or $\frac{2}{5}$

2 (a) 56 miles
 (b) 136 miles
 (c) (i) 240 km (ii) 130 km
 (d) About 170 miles

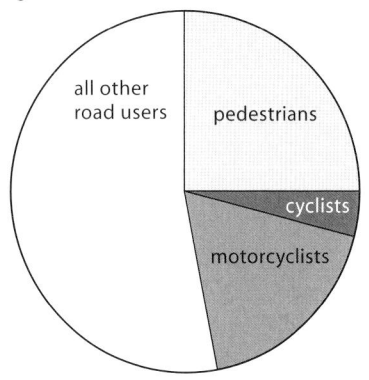

3 (a) $\frac{1}{4}$
 (b)

pie chart with sections: all other road users, pedestrians, cyclists, motorcyclists

4 (a) A full size drawing of

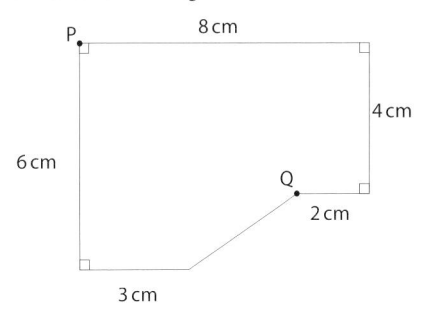

rectangle/shape with P at top-left, 8 cm top edge, 4 cm right edge, Q marked with 2 cm, 3 cm bottom, 6 cm left

 (b) 7.2 cm
 (c) 14.4 m

Index

3-D object
    net 57–58
    plan and elevations 58
12-hour and 24-hour clock 63

acute angle 69
adding without calculator
    decimals 81, 83
    integers 24–25
angle 68–73
    estimating 73
    measuring 68–70, 72
    types 68–71
area
    finding and estimating on centimetre grid 98–99
    of rectangle 100–101, 104–106
    of right-angled triangle 101, 106
    of shape made from rectangles 102–104
arrangements, listing 26–27

bar chart
    dual 34–35, 177
    for categorical data 32–33
    for numerical data 36–37
brackets in calculation 137

calculator, for square and square root 80
capacity 148–149
circle 15–16
circumference 15
clockwise, anticlockwise turn 71, 130–131
combined choices, listing 27–28
comparison of two data sets 51–52, 124–125
compass direction 130–131
composite shape, area of 102–104
cone 56
conversion between metric units 60–61, 75–77, 148–149
conversion graph 179–182
coordinates
    negative 120–121, 153
    positive 17–18, 150–156
cube 56–58
cuboid 56–57
    volume 135
cylinder 56

decimal
    1 place, locating on scale 42, 93–94
    1 place, ordering 42–43
    2 places, locating on scale 44, 93–94
    2 places, ordering 44–45
    adding without calculator 81, 83
    dividing by integer without calculator 97
    for capacity 149
    for length 46–47, 75–77
    for mass 61
    in area calculation 104–106
    multiplying by integer without calculator 96
    relating to fraction 161–162
    relating to percentage 162
    subtracting without calculator 82–83
diameter 15–16
directions, giving from street plan 130–131
distance table 166–167
dividing
    decimal by integer without calculator 97
    integer and decimal by 10, 100, … 54
    integer by integer without calculator 30
    mentally by 4, 5 84
divisibility 8–10
dual bar chart 34–35, 177

edge of solid 56
elevation of solid 58
enlargement by integer scale factor 114–116
equation, linear 126–129
equilateral triangle 17, 72
estimating
    angle 73
    answer to money problem 145–146
    area on grid of squares 99
    distance from drawn map scale 130–131
    length 89–91
    value from graduated scale 92–94
even number 8
expression, substituting into 136, 138–139, 142

face of solid 56–57
factor 11
formula
    in words 140–141
    substituting into 138–142

fraction
    of a number, finding 40–41
    relating to decimal 161–162
    relating to percentage 159–160, 162
    relating to pictorial representation 39–40
frequency chart
    for categorical data 32–33
    for numerical data 36–37

graph
    conversion 179–182
    linear, drawing from table of values 150–156
grid reference (letter–number) on map 130–131

hexagon 20–22

isosceles triangle 16–17, 72

kite 19

length 75–77
    estimating 89–91
linear function, completing table from context and
    drawing graph 150–156
linear sequence 86–87
listing arrangements and combined choices 26–28

map, interpreting 130–131
mass 60–61
mean 122–125
measuring
    angle 68–70, 72
    length 75
median 48–52
metric measures 149
mode 32–33, 36–37
money problem
    estimating answer to 145–146
    with calculator 146–147
    without calculator 144–145
multiple 10
multiplying
    by number ending in zeros 54–55
    decimal by integer without calculator 96
    integer and decimal by 10, 100, … 53
    integers without calculator 29
    mentally by 4, 5 84

negative coordinates 120–121, 153
negative numbers
    in difference calculations 119–120
    ordering 117–118
net of solid 57–58
number pattern 86–88

obtuse angle 69
octagon 21
odd number 8
operations, priority of 137
ordering
    decimals, 1 place 42–43
    decimals, 2 places 44–45
    integers 13
    negative numbers 117–118

parallelogram 19, 72
pentagon 20–22
percentage
    divided bar 173
    dual bar chart 177
    of a quantity, mentally 163–164
    pie chart 174–176
    relating to decimal 162
    relating to fraction 159–160, 162
perimeter 99, 103–104, 116
pictogram 35
pie chart 174–176
place value
    decimal 42–45, 53–54
    integer 12–13, 53–54
plan and elevations of solid 58
priority of operations 137
prism 56–58
probability
    from equally likely outcomes 110–111
    scale 108–109
pyramid 56–58

quadrilateral, special types 18

radius 15–16
range 50–52, 124–125
rectangle 18
    area 100–101, 104–106
reflection symmetry 17–19, 21–22

222

reflex angle 70
remainder, interpretation of 30, 145–146
rhombus 18
right angle 68
right-angled triangle 72
    area 101, 106
rounding to nearest 10, 100, … 13–14

scale drawing 46–47, 168–171
scale factor, integer, of enlargement 115–116
scale, graduated
    decimal 42, 44, 93–94
    reading and estimating integer value from 92–93
scalene triangle 16
sequence 86–88
solid
    net 57–58
    plan and elevations 58
sphere 56
square (shape) 18
square and square root 78–80
    notation 79
    on calculator 80
street plan, interpreting 130–131
substituting
    into expression or formula 136, 138–139, 142
    into formula in words 140–141

subtracting without calculator
    decimals 82–83
    integers 24–25
symmetry, reflection 17–19, 21–22

table of values, completing from context and using to draw linear graph 150–156
tally table, chart 32–33, 36–37
temperature 117–120
time 63–67
time interval calculations 64–65
timetable 66–67
trapezium 19
triangle
    equilateral 17, 72
    isosceles 16–17, 72
    right-angled 72
    scalene 16
two-way table 165–167

vertex of solid 56–57
views of solid 58
volume
    finding by counting centimetre cubes 134–135
    of cuboid 135
    of liquid 148–149

weight 60–61